ANDREW WALLACE is Professor of Mathematics at the University of Pennsylvania. He received his M. A. from the University of Edinburgh in 1946 and in 1949, received his Ph.D. from St. Andrews University where he was lecturer from 1949 to 1953. From 1950 to 1952, Dr. Wallace was on leave of absence as Commonwealth Fund Fellow at the University of Chicago. He was Assistant Professor at the University of Toronto from 1957 to 1959 and Professor at Indiana University from 1959 until 1964. Dr. Wallace's main research interests have been algebraic geometry and topology.

Differential Topology

First Steps

MATHEMATICS MONOGRAPH SERIES

Frederick J. Almgren, Jr., *Princeton University*
PLATEAU'S PROBLEM: AN INVITATION TO VARIFOLD GEOMETRY

John Lamperti, *Dartmouth College*
PROBABILITY: A SURVEY OF THE MATHEMATICAL THEORY

Kenneth Miller, *Columbia University*
LINEAR DIFFERENCE EQUATIONS

Robert T. Seeley, *Brandeis University*
AN INTRODUCTION TO FOURIER SERIES AND INTEGRALS

Michael Spivak, *Brandeis University*
CALCULUS ON MANIFOLDS: A MODERN APPROACH TO CLASSICAL THEOREMS OF ADVANCED CALCULUS

Andrew H. Wallace, *University of Pennsylvania*
DIFFERENTIAL TOPOLOGY: FIRST STEPS

Andrew H. Wallace
University of Pennsylvania

Differential Topology

First Steps

W. A. BENJAMIN, INC.
New York *Amsterdam*
1968

Differential Topology: First Steps

Library of Congress Catalog Card Number 67-30456
Manufactured in the United States of America

The manuscript was put into production on May 22, 1967;
this volume was published on February 1, 1968

W. A. BENJAMIN, INC.
New York, New York, 10016

Editors' Foreword

Mathematics has been expanding in all directions at a fabulous rate during the past half century. New fields have emerged, the diffusion into other disciplines has proceeded apace, and our knowledge of the classical areas has grown ever more profound. At the same time, one of the most striking trends in modern mathematics is the constantly increasing interrelationship between its various branches. Thus the present-day students of mathematics are faced with an immense mountain of material. In addition to the traditional areas of mathematics as presented in the traditional manner—and these presentations do abound—there are the new and often enlightening ways of looking at these traditional areas, and also the vast new areas teeming with potentialities. Much of this new material is scattered indigestibly throughout the research journals, and frequently coherently organized only in the minds or unpublished notes of the working mathematicians. And students desperately need to learn more and more of this material.

This series of brief topical books has been conceived as a possible means to tackle and hopefully to alleviate some of these pedagogical problems. They are being written by active research mathematicians, who can look at the latest developments, who can use these developments to clarify and condense the required material, who know what ideas to underscore and what techniques to stress. We hope that these books will also serve to present to the able undergraduate an introduction to contemporary research and problems in mathematics, and that they will be sufficiently informal that the personal tastes and attitudes of the leaders in modern mathematics will shine through clearly to the readers.

Topology is one of the branches of mathematics characteristic of this century. The undergraduate curriculum from calculus onwards has been deeply influenced by the development of general topology; and it is apparent that the techniques of classical algebraic topology have recently begun to have an effect as well. In just the past decade or so a new twig has sprouted from the general trunk of topology, and has grown into one of the most active and exciting branches of current mathematical research: differential topology. While it will undoubtedly be some time before this field comes to have vast general influence, it does have all the freshness and appeal of a new subject, and the student of mathematics will undoubtedly be curious to know something about it. Professor Wallace has provided in this book an introduction to differential topology for nonspecialists. The problems and techniques of differential topology are illustrated by many special cases and examples, so that readers with very little topological background will find the subject easily accessible; the book ends with a survey of further reading to guide the student whose interest has been aroused to a more detailed study.

Robert Gunning
Hugo Rossi

Princeton, New Jersey
Waltham, Massachusetts
October 1967

Preface

What is differential topology about? If this question were asked by a sufficiently advanced student with a good background in algebraic topology, it would be possible to give a fairly comprehensive answer. But it would be a technical answer. This book aims at giving an answer to a student at a much earlier stage in his career. The idea is to stimulate some intuitive feeling for certain aspects of the subject, while keeping the mathematical prerequisites to a minimum and avoiding the more difficult subtleties and technicalities.

Attention will be confined to the method of spherical modifications and the study of critical points of functions on manifolds. On the one hand, these ideas lend themselves to simple geometric description, while on the other, they have proved to be powerful tools in the study of the structure of manifolds. A simple illustration of their application is given in Chapter 7, namely, to the classification of two-dimensional manifolds.

To go further with the study of manifolds, which is the principal aim of differential topology, the geometric tools described here must be supplemented by more powerful algebraic tools. Some indication of the ideas required for such a study is given in Chapter 8.

In short, the first steps only are described here, and in this field, as indeed in any branch of topology, the first steps should be geometric, whereas the second or more technical steps should be based on an intuitive geometric feeling for the subject.

Knowledge of advanced calculus, including some properties of differential equations, is assumed, as is some knowledge of the behavior of quadratic forms under linear transformations of the variables. No previous knowledge of topology is required. All that is needed is described in the first chapter, but a student who has already been introduced to the ideas of open sets, closed sets, and continuous maps can safely go straight on to Chapter 2.

Chapters 2 and 3 introduce the reader to the notions of differentiable manifolds and maps. Then Chapter 4 discusses one of the central topics of differential topology, namely the theory of critical points of functions on a differentiable manifold. This study is continued into Chapter 5 with an investigation of level manifolds corresponding to a given function. This leads naturally to the definition in Chapter 6 of the idea of spherical modifications. In Chapter 7 the concepts of the previous chapters are applied to the classification problem of surfaces. Chapter 8 gives some guidance for further reading and study in this field.

Andrew H. Wallace

Philadelphia, Pennsylvania
November 1967

Contents

Set-Theoretic Symbols

The following is a vocabulary of set-theoretic symbols that are used in this book:

$x \in A$	x is a member of the set A
$A \subset B$	A is contained in B
$A \supset B$	A contains B
$A \cup B$	union of A and B
$A \cap B$	intersection of A and B
$\bigcup A_i$	union of the A_i
$\bigcap A_i$	intersection of the A_i
$\complement A$	complement of A
ϕ	empty set

1

Topological Spaces

1-1. NEIGHBORHOODS

General or point set topology can be thought of as the abstract study of the ideas of nearness and continuity. This is done in the first place by picking out in elementary geometry those properties of nearness that seem to be fundamental and taking them as axioms. Let E be n-dimensional Euclidean space and p a point in it. The idea of a neighborhood of p is that it should be a set of points near p and entirely surrounding p.

To make this precise, define a neighborhood of p to be any set U such that U contains an open solid sphere of center p. This makes the set U in Fig. 1-1 a neighborhood of p in the plane, since it contains an open disk with center p. But any disk with center p in Fig. 1-2 or 1-3 will contain points outside U, and so U is not a neighborhood of p in these cases. The definition of neighborhood is formulated in this way so as to be as free as possible from any ideas of size and shape, concepts that play no part in topology.

Using this definition of a neighborhood of a point in Euclidean space, it is easy to see that the following properties hold.

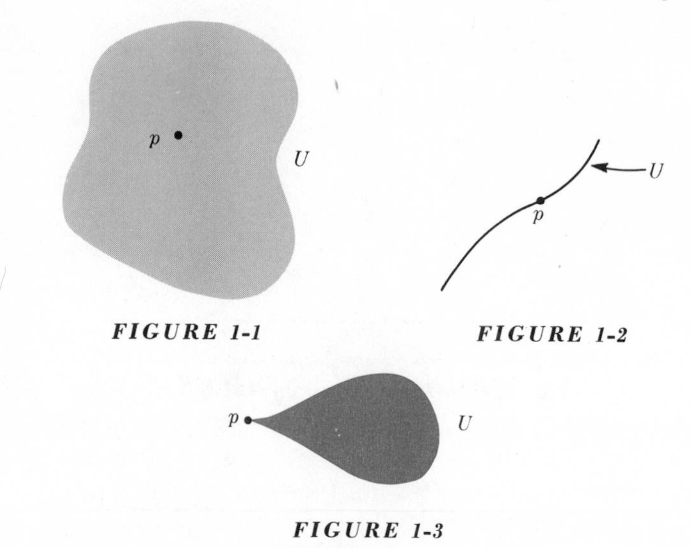

FIGURE 1-1

FIGURE 1-2

FIGURE 1-3

(1) p belongs to any neighborhood of p.

(2) If U is a neighborhood of p and $V \supset U$, then V is a neighborhood of p.

(3) If U and V are neighborhoods of p, so is $U \cap V$.

(4) If U is a neighborhood of p, then there is a neighborhood V of p such that $V \subset U$ and V is a neighborhood of each of its points.

Exercise. **1-1.** Prove properties 1 through 4.

A careful analysis of the properties of neighborhood and continuity, as appearing, for example, in the theorems of calculus, shows that they derive from the four properties listed above. It is therefore reasonable to take them as axioms in an abstract formulation. This leads to the following definition.

Definition 1-1. A *topological space* is a set E along with an assignment to each $p \in E$ of a collection of subsets of E, to be called *neighborhoods of p*, and satisfying the four properties listed earlier.

Examples

1-1. With neighborhoods as defined earlier, Euclidean space is a topological space.

1-2. Let S be the surface of a sphere, say the unit sphere in 3-space with center at the origin. Call U a neighborhood of p in S if, for some ϵ, U contains all the points of S at distance $<\epsilon$ from p. Verify that the axioms for neighborhoods are satisfied. Thus S is a topological space.

1-3. Other surfaces can be treated in the same way as the sphere. For example, the torus, the surface traced out by a circle of radius 1 and center (2, 0, 0) when the (x, y) plane is rotated about the y axis, is made into a topological space in this way. Also, spheres of higher dimensions become topological spaces in exactly the same way as the 2-sphere in Example 1-2.

Note that in Examples 1-2 and 1-3 the surrounding Euclidean space plays only an auxiliary role. The topological space in each case is a subset, and only points of that subset are of interest in defining the topology. In the same way any subset of a Euclidean space can be made into a topological space, and in fact the same thing can be done with a subset of any topological space. In more detail, let E be a topological space and let F be a subset. Let p be a point of F. Then a subset U of F will be called a *neighborhood of p in F* if $U = F \cap V$, where V is a neighborhood of p in E. As an exercise, check that the neighborhoods in F so defined satisfy the neighborhood axioms.

Definition 1-2. When F is made into a topological space by defining neighborhoods in this way, it is called a *subspace of E*.

Example

1-4. Examples 1-2 and 1-3 define the sphere and the torus as subspaces of Euclidean 3-space.

Note that all the examples of topological spaces given so far appear as subspaces of a Euclidean space. However, all topological spaces do not satisfy this property. For example, let E be the set of all bounded real-valued functions on the unit interval I of real numbers. Define U to be a neighborhood of p in E if U contains all q in E for which $\sup |p(x) - q(x)|$ $(x \in I)$ is less than some ϵ. It is easy to see that the neighborhood axioms are satisfied, but it can be shown (not so easily) that E is not a subspace of any Euclidean space.

Having made this remark, however, it can be forgotten so far as the reading of this book is concerned, for the spaces with which this book will deal will all be subspaces of Euclidean spaces.

1-2. OPEN AND CLOSED SETS

It turns out that there are two kinds of subsets of a topological space that are of particular importance.

Definition 1-3. If E is a topological space and U a subset, then U is called *open in E* (or simply *open*, if no confusion is likely) if, for each p in U, U is a neighborhood of p.

Definition 1-4. If E is a topological space and F is a subset, F is called *closed in E* (or simply *closed*) if $E - F$ is open.

Examples

1-5. Let E be the plane and let U be an open disk in E. Then U is an open set. Prove this as an exercise.

1-6. Let E be the plane and F a closed disk. Then F is a closed set.

1-7. Similarly, open and closed solid spheres of any dimension are open and closed sets of the corresponding Euclidean spaces.

1-8. In Euclidean n-space the set of points $(x_1, x_2, \ldots, x_n)$ satisfying inequalities $a_i < x_i < b_i$ $(i = 1, 2, \ldots, n)$ for fixed a_i and b_i is open. The set of points satisfying $a_i \leq x_i \leq b_i$ is closed.

The behavior of open and closed sets under the operations of union and intersection is of fundamental importance and is described by the following theorem.

Theorem 1-1. (1) *The union of any collection of open sets in a topological space is open.*

(2) *The intersection of a finite collection of open sets is open.*

(3) *The intersection of any collection of closed sets is closed.*

(4) *The union of a finite collection of closed sets is closed.*

Proof. (1) Let a collection of open sets in E be given, and denote the members of the collection by U_i where i ranges over some set of indices. Let $U = \cup U_i$ and take p in U. Then $p \in U_i$ for some i and so U_i is a neighborhood of p (Definition 1-3). But $U \supset U_i$ and so U is a neighborhood of p (neighborhood axiom (2)). Thus U is a neighborhood of each of its points and so, by Definition 1-3, it is open.

(2) Let U_1 and U_2 be open sets and take $p \in U_1 \cap U_2$. U_1 and U_2 are open and contain p, and so are neighborhoods of p (Definition 1-3). Hence $U_1 \cap U_2$ is a neighborhood of p (neighborhood axiom (3)). Thus $U_1 \cap U_2$ is a neighborhood of each of its points and thus is open (Definition 1-3).

Parts (3) and (4) of the theorem are obtained from parts (1) and (2) by taking complements.

Note that part (2) of the proof does not work for the intersection of an infinite number of open sets. For example, if E is the real line and U_n is the open interval $(-1/n, 1/n)$, each U_n is an open set but the intersection of all the U_n is the point 0, and this is not open.

Suppose now that A is any set in a topological space E. Theorem 1-1 says that the union Int A of all open sets contained in A is open. Clearly it is the "largest" open set contained in A.

Definition 1-5. Int A is called the *interior of* A.

Dually, the intersection $\bar{A}$ of all closed sets of E containing A is closed and is the "smallest" closed set containing A.

Definition 1-6. $\bar{A}$ is called the *closure of* A.

Definition 1-7. $\text{Fr } A = \bar{A} \cap \overline{\complement A}$ is called the *frontier of* A.

Example

1-9. Let E be the plane and let A be a disk including the points of the upper half circumference but excluding those of the lower half. Then Int A is the open disk, $\bar{A}$ is the closed disk, and Fr A is the circumference of the disk.

Exercises. **1-2.** Let A be a set in a topological space. Prove that a point p is in Int A if and only if p has a neighborhood contained in A. Prove, too, that p is in $\bar{A}$ if and only if every neighborhood of p meets A.

1-3. For any sets A and B in a topological space prove that $\overline{A \cup B} = \bar{A} \cup \bar{B}$ and $\overline{A \cap B} \subset \bar{A} \cap \bar{B}$.

1-4. Let E be a topological space and F a subspace. Prove that a set U in F is open in F if and only if $U = V \cap F$, where V is an open set in E.

1-3. CONTINUOUS MAPS

Let E and F be topological spaces and let f be a map of E into F. This is represented diagrammatically by the notation $f\colon E \to F$. The idea of continuity is simply that points that are near together in E are mapped into points that are near together in F. This is made precise as follows.

Definition 1-8. The map $f\colon E \to F$ is *continuous at* p if, given any neighborhood V of $f(p)$ in F, there is a neighborhood U of p in E such that $f(U) \subset V$. f is *continuous* if it is continuous at each p in E.

Exercises. **1-5.** Let E and F in Definition 1-8 both be the real line. The usual definition of continuity in this case is that f is continuous at x if, for any $\epsilon > 0$, there is a $\delta > 0$ such that $|f(x') - f(x)| < \epsilon$ whenever $|x' - x| < \delta$. Prove that this is equivalent to Definition 1-8.

1-6. Let $f\colon E \to F$ be a map. Prove that f is continuous if and only if the inverse image of any open set in F is open in E. Use this to show that the composition of continuous maps is continuous.

1-7. Let E be the union of two closed sets A and B and let $f\colon E \to F$ be a map. Suppose that the restrictions of f to A and B are continuous maps of A and B into F, respectively. Show that f is continuous. Give an example to show that this does not hold if A and B are not closed.

Continuous maps that have continuous inverses are of special importance.

Definition 1-9. Let f be a one-to-one map of E onto F. Thus there is an inverse map g of F onto E. If both f and its inverse are continuous, f will be called a *homeomorphism* and E and F will be said to be *homeomorphic*.

From the point of view of general topology, homeomorphic spaces are the same. That is to say, the properties that interest us are those that, when true for one space, are true for all spaces homeomorphic to it. Another way of looking at this is to note, first, that a homeomorphism between E and F sets up a one-to-one correspondence between the neighborhoods in E and the neighborhoods in F, and between the open sets in E and the open sets in F. Hence, any property defined entirely in terms of neighborhoods and open sets is a topological property. Some examples of such properties will appear presently.

1-4. TOPOLOGICAL PRODUCTS

This section describes a frequently used method of obtaining new spaces from given ones.

Let E and F be topological spaces. The set $E \times F$ is defined to be the set of pairs (p, q) where $p \in E$ and $q \in F$. This is made into a topological space as follows. If $(p, q) \in E \times F$, then a neighborhood of (p, q) is any set containing a set of the form $U \times V$, where U is a neighborhood of p in E and V is a neighborhood of q in F. It is not hard to see that the neighborhood axioms (1) through (4) are satisfied.

Definition 1-10. $E \times F$, made into a topological space as just described, is called the *topological product of E and F*.

Examples

1-10. If $E = F =$ the real line, then $E \times F$ is the plane with its usual topology as Euclidean 2-space.

1-11. If E is Euclidean 2-space and F is the real line, $E \times F$ is Euclidean 3-space. Clearly, this can be generalized; the topological product of Euclidean m-space and n-space is $(m + n)$-space.

1-12. If E is a real line interval and F is a circle, then $E \times F$ is a cylinder.

1-13. It will be seen later, in detail, that the torus is the topological product of a circle with itself.

1-5. CONNECTEDNESS

Two important topological properties will be described in this and the next section. The first, connectedness, is the property of being, so to speak, all in one piece.

Definition 1-11. A space E is *connected* if it cannot be expressed as the union of two nonempty disjoint sets open in E. A set in a topological space is *connected* if, as a subspace, it is a connected space.

Examples

1-14. Let E be the space consisting of two points a, b. The neighborhoods of a are to be the sets $\{a\}$ and $\{a, b\}$ and the neighborhoods of b are to be $\{b\}$ and $\{a, b\}$. It is easy to see that the neighborhoods axioms are satisfied. Also, the sets $\{a\}$ and $\{b\}$ are open; thus E is the union of two disjoint open sets. Hence, E is not connected.

1-15. Let A be the union of two disjoint open disks in the plane. Then A is not a connected set.

It is much more difficult to give an example of a space that is connected, except for something trivial, such as a space with only one point. One of the most important examples is the line interval. Intuitively it is fairly clear that this is all in one piece and so should be connected, but of course this needs proof.

Theorem 1-2. *Let A be an open line interval, say the set of real numbers x such that* $0 < x < 1$. *Then A is connected.*

Proof. Suppose that A is not connected. Then by definition $A = B \cup C$ where B and C are nonempty disjoint open sets on A, and so on the real line. Since B and C are not empty, there are points b in B and c in C. Suppose for the sake of definiteness that $b < c$. Then let D be the set of points x in B such that $x < c$. D is not empty, since it contains b. Let d be the least upper bound of numbers in D. The idea now is to show that d cannot belong to either B or C. Since, however, d is between b and c, it is certainly in A, and so in B or C. This contradiction will show that A is in fact connected.

Suppose, then, that $d \in B$. Since all x in D satisfy $x < c$, d satisfies $d \leq c$, and since d is in B, it must actually satisfy $d < c$. B is open and so there is an open interval U containing d and contained in B. If the length of U is taken less than $c - d$, U will then be contained in D. But then the right-hand end of U would be in D and would be greater than d, which is impossible since d is the upper bound of D. Hence, d is not in B.

Suppose that $d \in C$. Then, since C is open, there is an open interval U containing d and contained in C. But this means that, for a certain $\epsilon > 0$, there are no points of B, and so no points of D, between $d - \epsilon$ and d, and this contradicts the fact that d is the least upper bound of D. Hence, $d \notin C$.

Hence, d is neither in B nor in C, which leads to the required contradiction as explained earlier, thus completing the proof of the theorem.

It is clear that the same proof with minor modifications would show the connectedness of intervals with one or both end points included and of intervals that are infinite in one or both directions.

Theorem 1-2 has a converse that is fairly trivial.

Theorem 1-3. *If a set A of real numbers is connected, then A is an interval (finite or infinite, with or without end points).*

Proof. Let a and b be two points of A, with $a < b$. Then it is to be shown that A contains all numbers c such that $a < c < b$. Suppose that there is a c between a and b but not in A, and let B be the set of all numbers in A less than c, C the set of all numbers in A greater than c. Then $A = B \cup C$, B and C are both open in A and are nonempty and disjoint. This contradicts the assumed connectedness of A, and so there is no such c as assumed. Hence, A is an interval.

It now becomes fairly easy to construct other examples of connected spaces. The following theorems give general methods of getting new connected spaces from old ones.

Theorem 1-4. *Let $f: E \to F$ be a continuous map of a connected space onto a space F. Then F is connected.*

Proof. Suppose that the theorem is false. Then $F = A \cup B$ where A and B are nonempty, disjoint, and open. Then $E = f^{-1}(A) \cup f^{-1}(B)$, and $f^{-1}(A)$ and $f^{-1}(B)$ are disjoint and nonempty, and by Exercise 1-6 they are open. This contradicts the connectedness of E, and so F must be connected.

In particular, it follows from this that if E is connected and F is homeomorphic to E, then F is connected, so that connectedness is a topological property. Another example of the use of Theorem 1-4 may be had by noting that there is a continuous map of the unit interval of real numbers x such that $0 \leq x \leq 1$ onto the circumference of a circle. That is, map x on the point $(\cos 2\pi x, \sin 2\pi x)$ in the plane. Hence, the circumference of a circle is connected.

Theorem 1-5. *If E and F are connected spaces, then $E \times F$ is connected.*

Proof. As usual, the proof is by contradiction. Suppose that $E \times F$ is not connected. Then $E \times F = A \cup B$ where A and B are open, disjoint, and nonempty. Take (x, y) in A. The set $E \times \{y\}$ is homeomorphic to E and so is connected. It follows that $E \times \{y\}$ is contained in A. For otherwise its intersections with A and B would form a decomposition into open, disjoint, and nonempty sets. But then a similar reasoning would show that the slice $\{x'\} \times F$ would be in A for each x' in E, and so all of $E \times F$ would be in A, so that B would be empty. This gives the required contradiction, since B was assumed to be nonempty. Hence $E \times F$ is connected.

Example

1-16. It has already been seen that the closed line interval I is connected. It now follows that the rectangle $I^2 = I \times I$

is connected, and so by induction is the n-dimensional cube I^n. Similarly, since the real line is connected, n-dimensional Euclidean space is also connected.

Exercises. **1-8.** Let A be a connected set in a topological space E. Let B be a set such that $A \subset B \subset \bar{A}$. Then prove that B is connected.

1-9. Let A and B be connected sets in a space E and suppose that $A \cap \bar{B}$ is not empty. Prove that $A \cup B$ is connected.

Examples

1-17. Example 1-16 shows that an open disk (homeomorphic to I^2) is connected. Then Exercise 1-8 shows that a connected set is obtained by adding some or all of the points on the circumference. A similar example can be formulated for higher dimensions.

1-18. The surface of a sphere can be expressed as the union of two closed disks with nonempty intersection. So by Exercise 1-9 this surface is connected. Similarly, the n-sphere is connected for any n.

1-6. COMPACTNESS

The idea of compactness is a generalization of the property of being closed and bounded in a Euclidean space. First, the Hausdorff separation axiom will be assumed for the spaces to be considered. This assumption is not always made in general topology, but it is appropriate here since the spaces of most interest here will satisfy the condition anyway.

Definition 1-12. A topological space will be called *Hausdorff* if, for any two distinct points p and q, there are neighborhoods U of p and V of q such that $U \cap V = \varnothing$. Thus distinct points are separated by disjoint neighborhoods.

Examples

1-19. Any Euclidean space is Hausdorff.

1-20. Any subspace of a Euclidean space is Hausdorff. In fact, any subspace of any Hausdorff space is Hausdorff.

Before defining compactness, some preliminary definitions are needed.

Definition 1-13. A *covering* of a topological space E is a collection of sets in E whose union is E. It is called an *open covering* if all the sets of the collection are open.

Definition 1-14. Given a covering of a topological space, a *subcovering* is a covering whose sets all belong to the given covering.

Definition 1-15. A *compact space* is a Hausdorff space with the property that any open covering contains a finite subcovering, that is, a subcovering consisting of finitely many sets. A set in a topological space is *compact* if it is a compact subspace.

Examples

1-21. The Borel-Lebesgue theorem of analysis shows that a closed bounded set of a Euclidean n-space is compact (cf. [3]).

1-22. The real line is not compact. For take the collection of open intervals $(n - 1, n + 1)$, for all integers n. This is an open covering of the real line, but clearly no finite collection of these intervals can cover the whole line. A similar argument shows that Euclidean n-space is noncompact, and in fact so is any unbounded subset.

Exercises. **1-10.** Show that the n-sphere is compact for any n.

1-11. Prove that a closed subset of a compact space is compact, and that a compact set in any Hausdorff space is closed.

Note that a compact set in a Euclidean space must then be closed (Exercise 1-11) and bounded (Example 1-22). This gives the converse to the Borel-Lebesgue theorem. There is a general theorem that says that the topological product of compact spaces is compact (cf. [2]). This will not be proved here. However, one case will be wanted, namely, that in which the given compact spaces are subspaces of Euclidean spaces of dimensions m and n. The product is then a subspace of $(m + n)$-space. Since the given spaces are compact, they are closed and bounded, by the remark just made. Thus their product is closed and bounded (verify this!) in $(m + n)$-space and, hence, compact by the Borel-Lebesgue theorem.

Since compactness is defined in terms of open sets, it is a topological property. In fact, it is preserved by any continuous map.

Theorem 1-6. *Let $f: E \to F$ be a continuous map of a compact space E onto a Hausdorff space F. Then F is compact.*

Proof. Let an open covering of F be given, the individual sets being denoted by U_i with the i running over some set of indices. Then the sets $f^{-1}(U_i)$ form a covering of E that, by Exercise 1-6, is open. Since E is compact, a finite collection, say $f^{-1}(U_1), f^{-1}(U_2), \ldots, f^{-1}(U_n)$, will cover it. Then $U_1, U_2, \ldots, U_n$ forms a finite subcovering of the given covering of F. Since the Hausdorff condition has been assumed, F is then compact.

2

Differentiable Manifolds

2-1. INTRODUCTION

A number of the examples already given of topological spaces have the property that coordinates can be set up on them, at least locally around each point. In Euclidean n-space this is immediately evident, as a matter of definition. That is, each point actually is a set of n real numbers, namely, its coordinates. Consider on the other hand a two-dimensional sphere, say the unit sphere $x^2 + y^2 + z^2 = 1$, in 3-space. Take a point in the hemisphere $z > 0$. Here $z = (1 - x^2 - y^2)^{1/2}$, so that in fact the point is determined by the values of x and y. Thus (x, y) can be thought of as the coordinates of the point on the sphere. But they are only local coordinates in the sense that they determine points uniquely only in a certain open set, namely, the hemisphere $z > 0$. Note that the map taking the point (x, y, z) on the sphere onto the point $(x, y, 0)$ on the (x, y) plane is a homeomorphism of the hemisphere $z > 0$ onto the open unit disk, and we are actually using the coordinates of the image of a point under this map as the coordinates of the point on the hemisphere (Fig. 2-1).

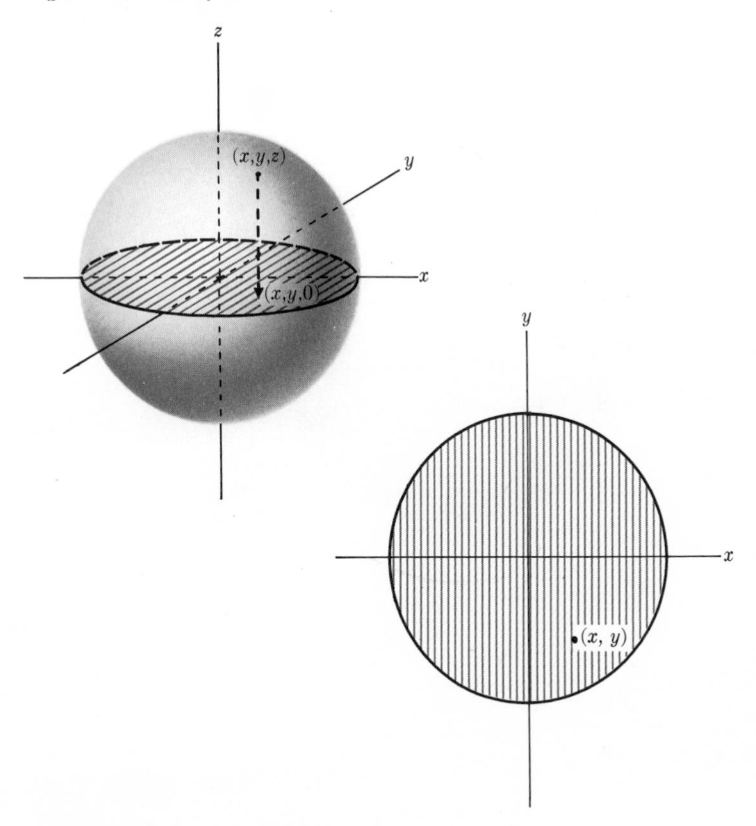

FIGURE 2-1 *Identification of upper hemisphere with disk in (x, y) plane by projection.*

The interesting feature of this example from our point of view is that the sphere can be covered by six hemispheres with similar properties, namely, the hemispheres $z > 0$, $z < 0$, $y > 0$, $y < 0$, $x > 0$, $x < 0$. Each hemisphere is mapped by a homeomorphism onto an open disk and the coordinates of points in the disk can be used as coordinates of points in the corresponding hemisphere. For example, in the hemisphere $x > 0$, (y, z) can be used as coordinates, and so on. The sphere is in this case said to be covered by six coordinate neighborhoods, that is, neighborhoods in which local coordinates can be set up.

A similar discussion can be carried out for the torus (see Fig. 2-2), although it is a bit more troublesome to do it all

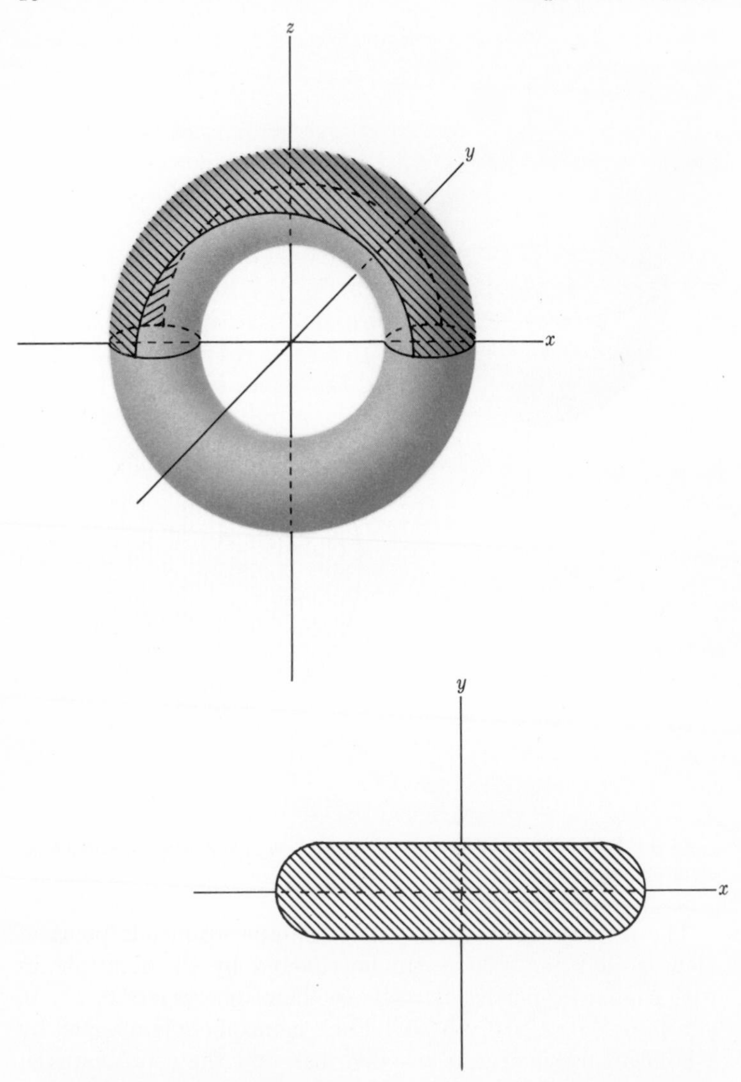

FIGURE 2-2 *Identification of top surface of torus (shaded dark) with shaded area of (x, y) plane.*

explicitly. Take the torus to be the surface generated by rotating the circle $(x - 2)^2 + y^2 = 1$ in the (x, y) plane about the y axis. Then, for example, (x, y) can be used as local coordinates on the surface in a neighborhood of the point $(0, 0, 3)$.

***Exercise.* 2-1.** Construct a complete set of coordinate neighborhoods covering the torus.

It will also be noticed that the foregoing examples have the property that if (x_1, x_2) and (y_1, y_2) are the local coordinates of a point in two overlapping coordinate neighborhoods, then y_1 and y_2 are differentiable functions of x_1 and x_2 and conversely. For example, in the case of the sphere consider the coordinate systems as above in the hemispheres $z > 0$ and $x > 0$. Let a point have coordinates (x_1, x_2) in the first of these neighborhoods and (y_1, y_2) in the second. Thus, if its coordinates in the surrounding Euclidean space are (x, y, z), this means that $x_1 = x$, $x_2 = y$, $y_1 = y$, $y_2 = z$. Since $x^2 + y^2 + z^2 = 1$ on the sphere,

$$\begin{aligned} y_1 &= x_2, \\ y_2 &= (1 - x_1^2 - x_2^2)^{1/2}. \end{aligned}$$

Clearly, the functions on the right of these equations have partial derivatives everywhere in the overlap of the two hemispheres. Similarly

$$\begin{aligned} x_1 &= (1 - y_1^2 - y_2^2)^{1/2}, \\ x_2 &= y_1, \end{aligned}$$

and again the functions on the right can be differentiated in the set $z > 0$, $x > 0$. It can also be checked easily that in this set the determinant

$$\begin{vmatrix} \dfrac{\partial y_1}{\partial x_1} & \dfrac{\partial y_1}{\partial x_2} \\ \dfrac{\partial y_2}{\partial x_1} & \dfrac{\partial y_2}{\partial x_2} \end{vmatrix}$$

is not zero.

The foregoing examples, particularly the features just described of the local coordinate systems on the sphere, motivate the definition of the type of space of special interest here, namely, the differentiable manifold. First, some preliminary definitions will be given.

2-2. DIFFERENTIABLE FUNCTIONS AND MAPS

Definition 2-1. Let U be an open set in Euclidean n-space E and let f be a real-valued function defined on U. Then f will be called *differentiable* if it has continuous partial derivatives of all orders with respect to the coordinates in E at all points of U.

Examples

2-1. A polynomial in the coordinates in E is differentiable on any open set in E. Of course, in this case all the derivatives of sufficiently high order vanish.

2-2. In Euclidean 2-space the function $(1 - x^2 - y^2)^{1/2}$ is not differentiable on any open set containing a point of the circle $x^2 + y^2 = 1$, but it is differentiable on any open set not containing any point of this circle.

2-3. Consider the function f on the real line defined as follows:

$$f(x) = \exp\left(\frac{1}{x^2 - 1}\right) \qquad \text{when } -1 < x < 1,$$
$$f(x) = 0 \qquad \text{if } x \leq -1 \text{ or } x \geq 1.$$

It should be checked as an exercise that this function has derivatives of all orders at all points of the line. Thus f is a differentiable function of the whole line. (Remember that $(1/t^n)e^{-1/t} \to 0$ as $t \to 0$ from above.)

2-4. The last example can be used to construct others in spaces of higher dimension. In Euclidean n-space let $r^2 = \Sigma x_i^2$ be the square of the distance from the origin, and define a function f as follows:

$$f(p) = \exp\left(\frac{1}{r^2 - 1}\right) \qquad \text{when } r < 1,$$
$$f(p) = 0 \qquad \text{when } r \geq 1.$$

Again this is a differentiable function on all of n-space.

Note that the function just constructed has the properties of being differentiable in all n-space, vanishing outside an open set (the unit solid sphere), and being nonzero inside this set. Functions with similar properties will be useful later.

Exercise. **2-2.** Use a method similar to that of the last example to construct a function differentiable on all n-space, equal to 1 on the closed solid sphere $\sum x_i^2 \leq 1$, and vanishing outside the solid sphere $\sum x_i^2 < 4$.

Since functions are sometimes given on sets that are not open, the following definition is sometimes needed to complement Definition 2-1.

Definition 2-2. Let f be a real-valued function on a set A in Euclidean n-space E. f will be called *differentiable on* A if it can be extended to an open set U containing A so that it is differentiable on U.

Thus, for example, to say that a function is differentiable at a point means that it is differentiable on a neighborhood of the point. The following is a natural extension of the idea of differentiability to maps between Euclidean spaces.

Definition 2-3. Let A be a set in Euclidean m-space and let $f_1, f_2, \ldots, f_n$ be n differentiable functions on A. Define a map f of A into Euclidean n-space by making $f(x)$ the point with coordinates $(f_1(x), f_2(x), \ldots, f_n(x))$, for x in A. Then f is called a *differentiable map of* A *into* n*-space.*

Note that if $n = 1$, the differentiable map simply becomes a differentiable function as in Definition 2-2.

In the study of differentiable maps an important part is played by the matrix of first partial derivatives of the f_i, in the notation of Definition 2-3.

Definition 2-4. The matrix with $\partial f_i/\partial x_j$ in the ith row and jth column is called the *Jacobian matrix* of the differentiable map f. If $n = m$, its determinant is called the *Jacobian determinant* of the map.

Exercises. **2-3.** Let f be a differentiable map of a set A in Euclidean m-space into n-space and let g be a differentiable map of $f(A)$ into p-space.

Prove that the composition gf is a differentiable map of A into p-space. Also, if F is the Jacobian matrix of f evaluated at a point x and if G is the Jacobian matrix of g evaluated at $f(x)$, prove that the Jacobian matrix of gf evaluated at x is GF.

2-4. Let f be a differentiable map of an open set in Euclidean n-space into n-space and suppose that f has a differentiable inverse. Show that the Jacobian determinant of f is nonzero at all points of U.

The result of Exercise 2-4 has a partial converse, namely, the inverse function theorem, which can be stated as follows.

Theorem 2-1. *Let f be a differentiable map of an open set U in n-space into n-space and let p be a point of U. Suppose that the Jacobian determinant of f is nonzero at p. Then there is a neighborhood V of p and a neighborhood W of $f(p)$ such that f maps V homeomorphically on W and the inverse of f on W is a differentiable map onto V.*

For the proof of this result see [3] or [6].

Note that this theorem gives only a local inversion of f. It is not possible in general to say anything more. For example, let U be the (x, y) plane with the origin removed and, using complex variable notation $z = x + iy$, define a map f of U onto itself by setting $f(z) = \exp(z)$. If the real coordinates of $f(z)$ are written as (u, v), then this map can be expressed in terms of real functions by the equations

$$u = \exp(x) \cos y,$$
$$v = \exp(x) \sin y.$$

It is easy to see that this is a differentiable map of U onto itself, and that the Jacobian determinant is nonzero on all of U. Certainly, then, f can be inverted locally, by Theorem 2-1. In fact, the inverse is given by $z = \log(u + iv)$ on any set that does not surround the origin. But f is not one-to-one on all of U, for if y is changed by adding to it any integral multiple of 2π, the value of $f(z)$ is unchanged.

2-3. DIFFERENTIABLE MANIFOLDS

With the preparation of the last section it is now possible to formulate the definition of the kind of space of special interest here, namely, the differentiable manifold.

Definition 2-5. *An n-dimensional differentiable manifold M* is a Hausdorff topological space that has a covering by countably many open sets $U_1, U_2, \ldots$, satisfying the following conditions:

(1) For each U_i there is a homeomorphism $\phi_i\colon U_i \to V_i$ where V_i is an open cell in Euclidean n-space.

(2) If $U_i \cap U_j \neq \varnothing$, the homeomorphisms ϕ_i and ϕ_j combine to give a homeomorphism $\phi_{ji} = \phi_j\phi_i^{-1}$ of $\phi_i(U_i \cap U_j)$ onto $\phi_j(U_i \cap U_j)$, which is a differentiable map.

Note that, in condition (2), ϕ_{ji} automatically has an inverse, namely ϕ_{ij}, and so its Jacobian determinant will be nonzero at all points of $\phi_i(U_i \cap U_j)$. Incidentally, the countability condition included in this definition will not play any part in the present discussion. Its inclusion in general is designed to avoid some pathological cases.

Examples

2-5. Euclidean n-space itself satisfies the conditions of this definition in a rather trivial way. Namely, the open covering can be taken to consist of one set only, U_1, the whole space; V_1 can also be taken as the whole n-space and ϕ_1 as the identity map.

2-6. Referring back to Section 2-1, we can see that the two-dimensional sphere S^2 satisfies the conditions of Definition 2-5. That is, S^2 is covered by six open sets, the various hemispheres described there, that play the part of the U_i. If U_1, for example, is the hemisphere $z > 0$, then the map ϕ_1 is to map the point (x, y, z) of U on the point $(x, y, 0)$; V_1 is being taken as the unit disk with the origin as center on the (x, y) plane. Similarly, if U_2 is the hemisphere $x > 0$, then $\phi_2(x, y, z) = (0, y, z)$. The map ϕ_{12} is then given by $\phi_{12}(0, y, z) = ((1 - y^2 - z^2)^{1/2}, y, 0)$ and this map is differentiable. A similar discussion can be carried out for the other ϕ_{ij}.

Exercises. **2-5.** Imitating the discussion of the last example, show that the n-dimensional sphere S^n is an n-dimensional differentiable manifold.

2-6. The projective plane P^2 is obtained from the 2-sphere S^2 by identifying diametrically opposite points. Thus a point of P^2 is a pair of antipodal points on S^2. If p is a point of P defined by the pair p_1, p_2 on S, take an open disk U_1 on S^2 around p_1 and the disk U_2 around p_2 whose points are opposite those of U_1. Then the points pairs contained in the pair of neighborhoods U_1 and U_2 will be called a neighborhood of p in P^2. The topology of P^2 will be defined by taking as a neighborhood of p any set containing a neighborhood of the type just described. Prove that P^2 is a two-dimensional differentiable manifold.

2-7. Generalize the last exercise by defining projective n-space P^n as the space obtained by identifying diametrically opposite points on the n-sphere. The neighborhoods are to be defined as in Exercise 2-6. Prove that P^n is a differentiable manifold.

2-8. Let X be the set of all $(n + 1)$-tuples of real numbers, excluding the one consisting of $n + 1$ zeros. Define $(x_0, x_1, \ldots, x_n) \sim (cx_0, cx_1, \ldots, cx_n)$ where c is any real number except zero. Prove that $\sim$ is an equivalence relation on X, and show that the equivalence classes are in one-to-one correspondence with the points of the projective space P^n as constructed in Exercise 2-7. Let U_i be the set of points of P^n with representatives in X for which $x_i \neq 0$. Such points then have representatives of the form $(x_0, x_1, \ldots, 1, \ldots, x_n)$ with 1 in the ith position. Let ϕ_i be the map of U_i into n-space carrying the point with representative $(x_0, x_1, \ldots, 1, \ldots, x_n)$ into $(x_0, x_1, \ldots, x_{i-1}, x_{i+1}, \ldots, x_n)$. Show that the U_i and ϕ_i so obtained can be taken as in Definition 2-5 to give P^n as a differentiable manifold.

2-9. Exercise 2-8 suggests the following construction. Let Z be the set of $(n + 1)$-tuples $(z_0, z_1, \ldots, z_n)$ of complex numbers excluding the zero $(n + 1)$-tuple $(0, 0, \ldots, 0)$, and write $(z_0, z_1, \ldots, z_n) \sim (cz_0, cz_1, \ldots, cz_n)$ where c is any nonzero complex number. Check that $\sim$ is an equivalence relation on Z, and let PC^n be the set of equivalence classes. Let U_i be the set of elements of PC^n with representatives of the form $(z_0, z_1, \ldots, 1, \ldots, z_n)$ with 1 in the ith position and let ϕ_i map an element of this form on $(z_0, z_1, \ldots, z_{i-1}, z_{i+1}, \ldots, z_n)$ in complex n-space, which is topologically a Euclidean $2n$-space. Show that the topology in PC^n can be defined so that the ϕ_i are all homeomorphisms, and that when this is done the U_i and ϕ_i define PC^n as a $2n$-dimensional differentiable manifold. PC^n is called complex projective n-space.

2-10. Prove that the torus (see Section 2-1) is a two-dimensional differentiable manifold.

2-4. LOCAL COORDINATES AND DIFFERENTIABLE FUNCTIONS

Definition 2-5 contains implicitly the idea behind the next definition. In effect, the homeomorphism ϕ_i identifies U_i with

V_i so that the Euclidean coordinates in V_i can be used to label the corresponding points in U_i. Thus ϕ_i can be thought of as establishing a coordinate system in U, or at least a local coordinate system, the word local signifying that the coordinates are set up only in an open set U_i and not over the whole manifold. This suggests that we should be able to define a notion of differentiable function on the manifold, using differentiation with respect to the local coordinates set up by the ϕ_i. The main point is to check that, in defining such a concept, the definitions in terms of different ϕ_i cannot contradict one another.

To check this point, let f be a real-valued function defined on an open set U of the differentiable manifold M. Take a point p in U and suppose it is in the set U_i of the open covering used in defining M. Consider the composed function $f\phi_i$ defined on $\phi_i(U \cap U_i)$. This function can be thought of as f expressed in terms of the local coordinates set up in U_i by the map ϕ_i. If p is also in another of the defining open sets U_j, then in a similar way $f\phi_j^{-1}$ is defined in a neighborhood of $\phi_j(p)$.

Lemma 2-1. *$f\phi_i^{-1}$ is differentiable in a neighborhood of $\phi_j(p)$ if and only if $f\phi_j^{-1}$ is differentiable in a neighborhood of $\phi_j(p)$.*

Proof. This follows at once, since $f\phi_i^{-1}$ can be written as $f\phi_j^{-1}\phi_j\phi_i^{-1} = f\phi_j^{-1}\phi_{ji}$. If $f\phi_j^{-1}$ is differentiable, then so is $f\phi_i^{-1} = f\phi_j^{-1}\phi_{ji}$, being a composition of differentiable functions (Exercise 2-3).

Thus this lemma says that if f, expressed in terms of the local coordinates set up by ϕ_i, is differentiable, then it is differentiable in terms of the local coordinates set up by ϕ_j, and conversely.

Definition 2-6. With the foregoing notation the function f will be said to be *differentiable in a neighborhood of* p if and only if $f\phi_i^{-1}$ is differentiable in a neighborhood of $\phi_i(p)$.

The lemma says that the condition stated here is independent of the ϕ_i that is used.

The function f will be called *differentiable on an open set* U if it is differentiable in a neighborhood of each point of U.

It will be called *differentiable on an arbitrary set* A if it is differentiable on an open set containing A.

Exercises. **2-11.** Let M be the sphere $x^2 + y^2 + z^2 = 1$ in 3-space. Prove that each of the Euclidean coordinates x, y, z is a differentiable function on M.

2-12. Repeat Exercise 2-11, taking M as the torus in 3-space, as in Section 2-1.

2-13. Let M be a differentiable manifold, p a point on it. Show that there is a neighborhood U of p and a differentiable function f on M that is positive in U and is zero at all points outside U. (*Hint:* Take U as a local coordinate neighborhood, mapped on an open set V of Euclidean space. By reducing the size of U, if necessary, take V as an open solid sphere of center the origin and choose the scale so that its radius is 1. Then use Example 2-4.)

2-14. Let A be a compact set in a differentiable manifold M and let U be an open set containing A. By using a suitable finite covering of A by neighborhoods, as obtained in the last exercise, construct a differentiable function f on M that is positive on A and zero outside U.

By using the same set of neighborhoods and some more, obtain a second function g that is equal to f on A and is nonzero on a neighborhood of the closure of the set where $f \neq 0$. By taking the quotient f/g construct a differentiable function on M that is equal to 1 on A to 0 outside U, and that otherwise takes values between 0 and 1.

Note in particular that if V_i is one of the open cells of Definition 2-5, then each of the Euclidean coordinates in V_i is a differentiable function on the corresponding U_i in M. That is, in each of the U_i there is a set of n differentiable functions whose values at a point can be taken (locally in U_i) as coordinates of the point. This suggests extending the notion of local coordinates by taking as coordinates in some neighborhood (not necessarily one of the U) the values of n differentiable functions on that neighborhood. Of course, some condition will have to be imposed to ensure that these coordinates correspond to a homeomorphism of the neighborhood on a Euclidean cell.

So let $f_1, f_2, \ldots, f_n$ be n differentiable functions in some neighborhood of a point p of M and suppose that p is in U_i, in the notation of Definition 2-5. Thus $f_1\phi_i^{-1}, f_2\phi_i^{-1}, \ldots, f_n\phi_i^{-1}$ are differentiable functions in the sense of Definition 2-2 around $\phi_i(p)$ in V_i. Suppose that the Jacobian determinant of these functions with respect to the Euclidean coordinates in V_i is not zero at $\phi_i(p)$. Then map a neighborhood of $\phi_i(p)$ into Euclidean n-space by mapping a point x on the point with coordinates $(f_1\phi_i^{-1}(x), f_2\phi_i^{-1}(x), \ldots, f_n\phi_i^{-1}(x))$; call the resulting map f. According to Theorem 2-1 there is a neighborhood V of $f\phi_i(p)$ that can be taken as an open cell and a neighborhood U' of $\phi_i(p)$ such that f maps U' homeomorphically on V, the inverse again being a differentiable map. Finally, write $U = \phi_i^{-1}(U')$ and $\phi = f\phi_i$, so that ϕ maps a point q in U onto the point with coordinates $(f_1(q), f_2(q), \ldots, f_n(q))$ in V. Clearly, ϕ is a homeomorphism of U onto V.

Definition 2-7. The homeomorphism ϕ will be called a *local coordinate system on the neighborhood* U of p. U is called the corresponding *local coordinate neighborhood.*

The motivation for this definition is that points of U are named by the coordinates of their images under ϕ. Note that the ϕ_i and U_i of Definition 2-5 are automatically local coordinate systems in the sense of this definition.

Exercises. **2-15.** Show that the definition of local coordinates around a point is independent of the choice of ϕ_i used in the construction.

2-16. Let ϕ and ψ be any two local coordinate systems around p, mapping neighborhoods of p on open cells V and W, respectively, in Euclidean n-spaces. Show that $\phi\psi^{-1}$ is a differentiable map of an open set in W on an open set in V, with nonzero Jacobian determinant.

Note that the result of Exercise 2-16 means that we can add to the coordinate neighborhoods U_i defining a differentiable manifold (as in Definition 2-5) other coordinate neighborhoods, obtained as in Definition 2-7, and the combined collection of local coordinate systems will continue to satisfy condition (2) of Definition 2-5.

2-17. Let p be a point of a differentiable manifold M and let ϕ be a local coordinate system in a neighborhood U of p as in Definition 2-7. Let f be a differentiable function on a neighborhood of p (Definition 2-6). Prove that $f\phi^{-1}$ is a differentiable function on a neighborhood of $\phi(p)$ in V (in the notation of Definition 2-7).

The meaning of this result is that, while the notion of differentiable function is defined using only the coordinate neighborhoods U_i of the definition of a manifold, the same description of differentiability holds in terms of the additional local coordinate systems introduced by Definition 2-7.

2-18. Take the (x, y) plane as a differentiable manifold M. Show that polar coordinates (r, θ) defined by $x = r\cos\theta$, $y = r\sin\theta$ can be used as local coordinates in any open disk not containing the origin.

2-19. Similarly, show that the spherical polar coordinates (r, θ, ϕ) defined by $x = r\sin\phi\cos\theta$, $y = r\sin\phi\sin\theta$, $z = r\cos\phi$, can be used as local coordinates in 3-space in any open sphere not meeting the z axis.

2-20. The spherical polar coordinates of the last exercise can be restricted to the surface of the unit sphere by fixing $r = 1$. Show that on the surface of this sphere (θ, ϕ) can be used as local coordinates in any disk not containing the point with Euclidean coordinates $(0, 0, \pm 1)$.

2-5. DIFFERENTIABLE MAPS

It was seen in Section 2-2 that differentiable functions on an open set in Euclidean space can be put together to define the notion of a differentiable map. The same sort of thing is now to be done more generally to define differentiable maps of one manifold into another. These differentiable maps play the same sort of role in the theory of differentiable manifolds that continuous maps play in the theory of topological spaces in general. The idea of the following definition is to use local coordinate systems to transfer the already familiar Definition 2-3 to differentiable manifolds.

Definition 2-8. Let M, N be differentiable manifolds of dimensions m and n, respectively, and let U be an open set on M, and f a map of U into N. Let p be a point of U and let ϕ be a local coordinate system on N in a neighborhood W of $f(p)$; thus ϕ is a homeomorphism of W onto an open cell V in some Euclidean n-space. Thus ϕf becomes a map of a neighborhood of p into V, which can be described by writing the individual coordinates in V as functions on a neighborhood of p. If these functions are differentiable, the map f will be said to be *differentiable on a neighborhood of* p. f will be called *differentiable on* U if it is differentiable on a neighborhood of each point of U.

Note that Definition 2-8 could also be formulated as follows. Let ψ be a local coordinate system on a neighborhood U_0 of p, assumed to be contained in U, ψ being a homeomorphism on an open cell V_0 in Euclidean n-space. Then the composition $\phi f \psi^{-1}$ is a map of V_0 into V. Definition 2-8 says that if this map is differentiable, then f is differentiable on a neighborhood of p. Another point that should be noted (and checked as an exercise!) is that the definition of differentiability of a map does not depend on the coordinate system used around $f(p)$. That is, if the stated condition holds for one such local coordinate system, it holds for any other.

Definition 2-8 is extended to maps defined on arbitrary sets as follows.

Definition 2-9. Let M and N be differentiable manifolds, let A be a subset of M, and let f be a map of A into N. f will be called *differentiable* if it can be extended to a differentiable map on an open set U containing A.

Examples

2-7. If M and N are taken as Euclidean spaces, note that Definitions 2-8 and 2-9 reduce to Definition 2-3.

2-8. Any differentiable function (Definition 2-6) on a set U of a differentiable manifold M is a differentiable map into Euclidean 1-space. More generally, if $f_1, f_2, \ldots, f_r$ are differentiable functions on U, the map taking p into $(f_1(p), f_2(p), \ldots, f_r(p))$ is a differentiable map of U into Euclidean r-space.

Exercises. **2-21.** Let M be the torus in 3-space (see Section 2-1) and let (l, m, n) be the direction cosines of the outward normal to M at p. Then, since $l^2 + m^2 + n^2 = 1$, (l, m, n) can be taken as a point of the unit 2-sphere S^2 in 3-space. Call this point $f(p)$. Then prove that f is a differentiable map of M into S^2.

2-22. The position of a point q on the torus can be specified as follows, representing the torus as in Section 2-1. On the circle $(x - 2)^2 + y^2 = 1$ in the (x, y) plane, take the point whose direction from the center makes an angle θ with the positive direction of the x axis. Draw the circle through this point parallel to the (x, z) plane and lying on the torus, and take the point at angular distance ϕ round this circle (Fig. 2-3). Then the position of q can be specified by the pair (θ, ϕ). Note that these are not coordinates over all of the torus, since for a given q the angles θ and ϕ are defined only up to integral multiples of 2π. Show,

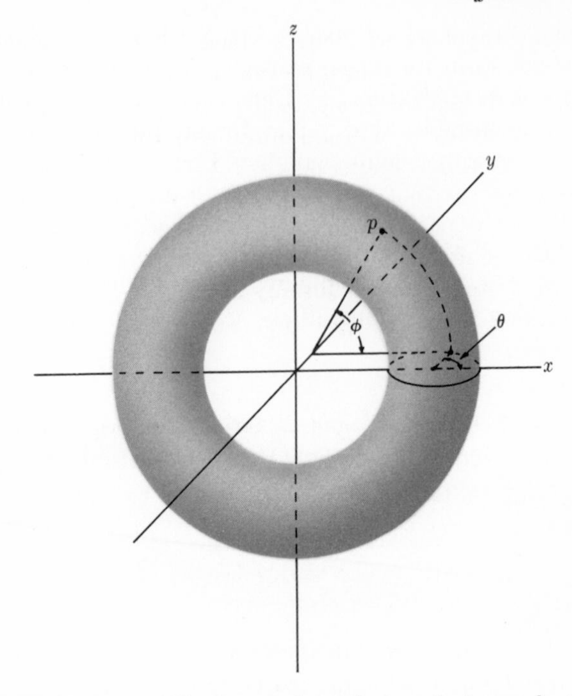

FIGURE 2-3 *Naming a point of the torus by two angles θ, ϕ.*

however, that (θ, ϕ) can be taken as local coordinates on suitable open sets of the torus. Now take p in the (x, y) plane with coordinates (x, y) and let $f(p)$ be the point on the torus with the angular coordinates (in the sense just explained) $\theta = x$, $\phi = y$. Prove that f is a differentiable map of Euclidean 2-space onto the torus.

Notice that, if the map f is restricted to the square $0 \leq x \leq 2\pi$, $0 \leq y \leq 2\pi$, then f is one-to-one on the interior and points on the sides opposite each other are mapped on the same point. This can be expressed pictorially by saying that the torus is obtained from the square by identifying the pairs of opposite sides.

Another interpretation of the exercise is obtained by thinking of θ and ϕ each as an angular coordinate on a circle. Thus the torus is represented as the topological product of two circles.

2-23. Let M, N, and P be differentiable manifolds and let $f: M \to N$ and $g: N \to P$ be differentiable maps. Show that the composition gf is a differentiable map.

2-24. Let p be a point of the 2-sphere S^2 and let $f(p)$ be the point of the projective plane P^2 (cf. Exercise 2-6) obtained by identifying p with the diametrically opposite point. Show that f is a differentiable map of S^2 onto P^2.

In general topology we identify spaces that are homeomorphic. Here, however, before we can regard two differentiable manifolds as being the same, a stronger condition must be satisfied, namely, the condition of diffeomorphism.

Definition 2-10. Let M and N be differentiable manifolds, and let f be a one-to-one differentiable map of M onto N such that the inverse map is also differentiable. Then f is called a *diffeomorphism* and the manifolds M and N are said to be *diffeomorphic.*

***Exercise.* 2-25.** Let f be a one-to-one differentiable map of M onto N. For any p on M, let ϕ be a local coordinate system in a neighborhood of p and let ψ be a local coordinate system in a neighborhood of $f(p)$ on N, so that $\psi f \phi^{-1}$ is a differentiable map of one n-cell on another. Suppose that, expressed in terms of the Euclidean coordinates, the Jacobian determinant of this map is nonzero in a neighborhood of $\phi(p)$; suppose that a similar condition holds in a neighborhood of every point of M. Prove that f is a diffeomorphism.

Note that this exercise gives an alternative way of formulating Definition 2-10. Note also that diffeomorphic differentiable manifolds are automatically of the same dimension.

2-6. RANK OF A DIFFERENTIABLE MAP

The definition to be given here is motivated by the observation that a differentiable map defined on a manifold may lead to a drop in dimension. For example, consider the map f of the plane into itself defined by mapping the point (x, y) on the point $(x, 0)$. Here f is defined on something of dimension two but its image is of dimension one. This observation will not be exploited in full until the next chapter, but in the meantime it should be noted that in the example just given the Jacobian determinant of the map is everywhere zero. On the other hand, if this determinant were everywhere nonzero, f would have a local inverse and thus it would be expected that its image would have dimension two. Thus the drop in dimension should be connected in some way with the rank of the Jacobian matrix of the map f.

Definition 2-11. Let M and N be differentiable manifolds and let $f: M \to N$ be a differentiable map. Let $p \in M$ and let $\phi: U \to V$ and $\psi: U' \to V'$ be local coordinate systems around p and $f(p)$, respectively. Thus $\psi f \phi^{-1}$ is a differentiable map of the Euclidean open set V into V'. Then the *rank of f at p* is defined as the rank of the Jacobian matrix of $\psi f \phi^{-1}$ at $\phi(p)$. If the rank at all points is r, then f will be said to be *of rank r*.

Exercises. **2-26.** Verify that Definition 2-11 is independent of the choices of local coordinates around p and $f(p)$.

2-27. Check that the map f of the plane into itself defined by $f(x, y) = (x, 0)$ has rank 1 everywhere.

2-28. Let $f: M \to N$ be a differentiable map of rank $m = \dim M \leq n = \dim N$. Show that, in terms of suitable local coordinates $(x_1, x_2, \ldots, x_m)$ around a point p and $(y_1, y_2, \ldots, y_n)$ around $f(p)$, the map f is given locally by

$$y_i = x_i \qquad (i = 1, 2, \ldots, m),$$
$$y_i = f_i(x_1, x_2, \ldots, x_m) \qquad (i = m + 1, \ldots, n),$$

where the f are differentiable functions of their arguments.

2-7. MANIFOLDS WITH BOUNDARY

There is an important extension of the idea of a differentiable manifold as defined in Section 2-3. This extension is motivated by the example of a closed disk. An interior point p of D has a neighborhood that is an open disk, but a point q on the boundary has a neighborhood (in the disk) that is a half disk. Thus, if D is to be thought of as a differentiable manifold, two different kinds of coordinate neighborhoods will have to be considered, according as the point in question is interior or not.

Definition 2-12. A *differentiable manifold of dimension n with boundary* is a topological space M with a subspace N and a countable open covering $U_1, U_2, \ldots$, with homeomorphisms $\phi_1, \phi_2, \ldots$, satisfying the following conditions.

(1) Each set U_i of the given covering is either contained in $M - N$, in which case there is a homeomorphism $\phi_i: U_i \to V_i$,

where V_i is a solid open sphere in n-space, or otherwise there is a homeomorphism ϕ_i: $U_i \to V_i$ where V_i is a hemisphere of the form $\Sigma_1^n x_i^2 < 1$, $x_n \geq 0$, the set $U_i \cap N$ being mapped on the subset of V_i for which $x_n = 0$.

(2) If U_i and U_j are two sets of the given covering and if ϕ_i and ϕ_j are the homeomorphisms just described and if $U_i \cap U_j \neq \phi$, then $\phi_{ij} = \phi_i\phi_j$ is a differentiable map of $\phi_j(U_i \cap U_j)$ onto $\phi_i(U_i \cap U_j)$.

Since homeomorphisms of the type ϕ_{ij} must carry interior points into interior points and frontier points into frontier points, it is clear that if those U_i that meet N are restricted to N, they form a covering that defines a structure of an $(n - 1)$-dimensional differentiable manifold on N.

Example

2-9. The solid sphere and the solid torus are differentiable 3-manifolds with boundaries, the boundaries, of course, being the sphere and the torus, respectively.

It is easy to see that the definitions given in the earlier sections of this chapter extend to differentiable manifolds with boundaries, with minor alterations to take care of the boundary points.

It can also be shown (but the proof is difficult and it will not be given here) that if two differentiable manifolds with boundaries have diffeomorphic boundaries, they can be put together to form a differentiable manifold by identifying the boundaries. In this way, for example, two disks can be fitted together to form a sphere. One disk becomes the top hemisphere, the other becomes the lower hemisphere, and the boundaries are identified to become the equator.

In more detail, this process can be described as follows. Let M_1 and M_2 be differentiable manifolds with boundaries N_1 and N_2, respectively, and let f: $N_1 \to N_2$ be a given diffeomorphism. Form a set M whose elements are the points of $M_1 - N_1$, the points of $M_2 - N_2$, and the points pairs $(p, f(p))$ with p in N_1. Thus M is the union of M_1 and M_2 with all the pairs $(p, f(p))$ identified as single points. The pair $(p, f(p))$

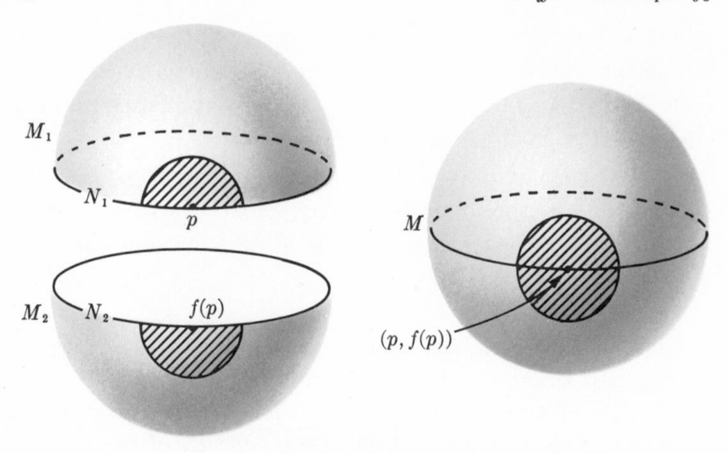

FIGURE 2-4

can thus be unambiguously called p. To make M into a topological space, neighborhoods of its points must be defined. If p is in $M_1 - N_1$ or $M_2 - N_2$, then its neighborhoods in M_1 or M_2, respectively, are to be taken as its neighborhoods in M. On the other hand, if $p = (p, f(p))$ is an identified pair with p in N_1 and $f(p)$ in N_2, then each neighborhood of p in M is to be the union of a neighborhood of p in M_1 and a neighborhood of $f(p)$ in M_2, with the appropriate identifications of points in N_1 and N_2.

Thus, for example, in Fig. 2-4 the two semicircular neighborhoods U_1 and U_2 of p and $f(p)$ combine to form a neighborhood U of $(p, f(p))$ in M.

The difficult step now is to show that M is a differentiable manifold. For the proof in detail, see [6], Section 6. What we must do is show that, if coordinate neighborhoods U_1 and U_2 fit together as just described to form a neighborhood U in M, and V_1 and V_2 fit together to form V, then the coordinate changing functions (functions such as the ϕ_{ij} of Definition 2-12) for $U_1 \cap V_1$ can be fitted together with those for $U_2 \cap V_2$ to give coordinate change functions for M that are differentiable.

3

Submanifolds

3-1. THE DEFINITION

In studying differentiable manifolds, the notion of subspace is too general. If one manifold is contained in another, it is desirable that the local coordinate systems in the two be related in some simple way. The following is the appropriate definition.

Definition 3-1. Let M be a differentiable manifold of dimension m and let N be a subset of M satisfying the conditions:

(1) N is a differentiable manifold of dimension n.

(2) If p is a point of N, there is a local coordinate neighborhood U of p in M with a local coordinate system $\phi: U \to V$, where V is an open cell in Euclidean m-space, such that $\phi(N \cap U)$ is the subset of V satisfying $x_{n+1} = x_{n+2} = \cdots = x_m = 0$, and ϕ restricted to $U \cap N$ is a local coordinate system for N around p.

Then N is called a *submanifold of* M.

Condition (2) can be stated a bit more informally by saying that local coordinates $x_1, x_2, \ldots, x_m$ are set up around p on M so that N has the local equations $x_{n+1} = x_{n+2} = \cdots = x_m = 0$, while $x_1, x_2, \ldots, x_n$ are local coordinates on N around p. Thus, local coordinate systems on N are induced by those on M. This condition also ensures that N is surrounded, in M, by a kind of tubular neighborhood. This point will be discussed later, but first some examples will be given.

Examples

3-1. The simplest example, of course, is obtained by taking M to be Euclidean m-space and N the Euclidean subspace given by the equations $x_{n+1} = x_{n+2} = \cdots = x_m = 0$.

3-2. Take M to be Euclidean 3-space. The coordinates x, y, z in 3-space can then be used as local coordinates around any point. Take N to be the 2-sphere with equation $x^2 + y^2 + z^2 = 1$. This has already been seen to be a differentiable manifold (Example 2-6). It will now be shown that condition (2) of Definition 3-1 is satisfied. Take, for example, a point p on the hemisphere $z > 0$ of N. It is evident that the coordinates (x, y, z) cannot be used as local coordinates around p in M if condition (2) is to be satisfied. However, new coordinates can be defined by the transformation

$$\begin{aligned} X &= x, \\ Y &= y, \\ Z &= z - (1 - x^2 - y^2)^{1/2}. \end{aligned}$$

The functions on the right are differentiable, as are the functions expressing the inverse transformation, and the Jacobian determinant of X, Y, Z with respect to x, y, z is nonzero around any point on the hemisphere $z > 0$. Thus (X, Y, Z) are admissible as local coordinates around p (Definition 2-7). In terms of these coordinates, N has the equation $Z = 0$ in a neighborhood of p. Also, it was seen in Example 2-6 that X and Y can be taken as local coordinates on N in the hemisphere $z > 0$. Thus condition (2) of Definition 3-1 holds around p. A similar argument shows that the condition holds around any point of N.

An important feature of this example is that it illustrates that given local coordinate systems may not make it obvious that a subset of a differentiable manifold is a submanifold. Some adjustment of the local coordinate systems may be needed before condition (2) can be seen to hold.

3-3. Take M to be the 2-sphere $x^2 + y^2 + z^2 = 1$ in 3-space and let N be the circle in which it intersects the plane $z = 0$. Since z can always be taken as one of the local coordinates on M at any point of N, the other being either x or y, it is easy to see that N is a submanifold of M. Clearly, this example can be generalized to higher dimensions.

3-4. It is of interest to look at an example in which condition (2) of Definition 3-1 is not satisfied. In Exercise 2-22 an example was given of a differentiable map ϕ of the plane E_2 onto the torus. Let L be a straight line in E passing through the origin and having irrational slope. Then it is easy to see that ϕ maps L in a one-to-one manner into the torus. Take the torus as M and $\phi(L)$ as N; L, a one-dimensional Euclidean space, is a differentiable manifold. Its image N, however, winds infinitely often round M, in fact, in such a way that if p is in N and U is any neighborhood of p in M, then $N \cap U$ will consist of infinitely many disjoint segments. Thus no choice of local coordinates around p in M can make condition (2) of Definition 3-1 work. Thus N is not a submanifold of M.

3-2. MANIFOLDS IN EUCLIDEAN SPACE

A case of special interest is that in which the containing manifold is a Euclidean space. So let M be a differentiable manifold of dimension n, given as a submanifold of Euclidean N-space E. Condition (2) of Definition 3-1 is expressed, in a neighborhood of some point p of M, in terms of some local coordinate system in E in a neighborhood of p. On the other hand, the Euclidean coordinates themselves can be used as local coordinates around p in E, so that it is natural to ask how condition (2) of Definition 3-1 appears when expressed in terms of the Euclidean coordinates. This question will be examined by making the necessary coordinate transformation in two stages. The result will appear as Theorem 3-1.

Start with a local coordinate system in a neighborhood U of a point p of M for which condition (2) of Definition 3-1 is satisfied. This is a homeomorphism $\phi\colon U \to V$ where V is an open cell in a second Euclidean N-space E' in which $y_1, y_2, \ldots, y_N$ will be used as coordinates. In particular, ϕ restricted to $U \cap M$ is a local coordinate system on M around p and it maps this set on the part of V for which $y_{n+1} = y_{n+2} = \cdots = y_N = 0$. On the other hand, the identity map of U on itself is a local coordinate system in U and so, according to condition (2) of Definition 3-1, ϕ^{-1} can be expressed by equations

$$x_i = \psi_i(y_1, y_2, \ldots, y_N), \qquad i = 1, 2, \ldots, N, \tag{1}$$

where the ψ_i are differentiable and the Jacobian determinant

$$\left| \frac{\partial(\psi_1, \psi_2, \ldots, \psi_N)}{\partial(y_1, y_2, \ldots, y_N)} \right|$$

is nonzero throughout V.

In particular this determinant is not zero at $\phi(p)$ and so, with suitable renumbering of the x, the determinant

$$\left| \frac{\partial(\psi_1, \psi_2, \ldots, \psi_n)}{\partial(y_1, y_2, \ldots, y_n)} \right|$$

is not zero at $\phi(p)$. So if the map θ of a neighborhood of $\phi(p)$ into a third Euclidean space E'', where the coordinates are $z_1, z_2, \ldots, z_N$, is defined by

$$\begin{aligned} z_1 &= \psi_1(y_1, y_2, \ldots, y_N) \\ &\;\;\vdots \\ z_n &= \psi_n(y_1, y_2, \ldots, y_N) \\ z_{n+1} &= y_{n+1} \\ &\;\;\vdots \\ z_N &= y_N, \end{aligned} \tag{2}$$

the Jacobian of the functions on the right is not zero at $\phi(p)$ and so is nonzero in a neighborhood of $\phi(p)$. It follows, by

Theorem 2-1, that θ maps a neighborhood V' of $\phi(p)$ homeomorphically on an open cell W in E'', and θ^{-1}, defined on W, is differentiable.

Now let $U' = \phi^{-1}(V')$ and write $\chi = \theta\phi$. χ is thus a homeomorphism of U' on W and, by combining the Eqs. (1) and (2) of ϕ^{-1} and θ, χ must have equations of the form

$$\begin{aligned} z_1 &= x_1 \\ &\;\;\vdots \\ z_n &= x_n \\ z_{n+1} &= \chi_{n+1}(x_1, x_2, \ldots, x_N) \\ &\;\;\vdots \\ z_N &= \chi_N(x_1, x_2, \ldots, x_N). \end{aligned} \tag{3}$$

Since χ is the composition of θ and ϕ, the functions on the right have nonzero Jacobian determinant in U' (cf. Exercise 2-3) and so χ is also a local coordinate system. The inverse χ^{-1} will be expressed by equations of the form

$$\begin{aligned} x_1 &= z_1 \\ &\;\;\vdots \\ x_n &= z_n \\ x_{n+1} &= \lambda_{n+1}(z_1, z_2, \ldots, z_N) \\ &\;\;\vdots \\ x_N &= \lambda_N(z_1, z_2, \ldots, z_N). \end{aligned} \tag{4}$$

Now in the first place $\chi(U' \cap M)$ is precisely the set in W satisfying the equations $z_{n+1} = z_{n+2} = \cdots = z_N = 0$, and so, using Eqs. (4), it follows that $U' \cap M$ is the set of points in U' satisfying

$$\begin{aligned} x_{n+1} &= \lambda_{n+1}(x_1, x_2, \ldots, x_n, 0, 0, \ldots, 0) \\ &\;\;\vdots \\ x_N &= \lambda_N(x_1, x_2, \ldots, x_n, 0, 0, \ldots, 0). \end{aligned} \tag{5}$$

And in the second place ϕ, restricted to $U' \cap M$, gives a local coordinate system on M and so the restriction of χ is also a local coordinate system in the neighborhood $U' \cap M$. But it is easy to see that χ maps $(x_1, x_2, \ldots, x_N)$ on $U' \cap M$ onto the point $(z_1, z_2, \ldots, z_n, 0, 0, \ldots, 0)$ with $x_i = z_i$ for $i = 1, 2, \ldots, n$. That is, the projection map of $U' \cap M$ into the space $x_{n+1} = x_{n+2} = \cdots = x_N = 0$ is itself a local coordinate system on $M \cap U'$.

This discussion can be summed up in the following theorem.

Theorem 3-1. *If M is an n-dimensional submanifold of Euclidean N-space E and if p is a point of M, then, with suitable numbering of the Euclidean coordinates $x_1, x_2, \ldots, x_N$ in E, the projection on the space $x_{n+1} = x_{n+2} = \cdots = x_N = 0$ is a local coordinate system on M in a neighborhood of p, while in a neighborhood of p in E, M is the set of points satisfying equations (obtained from (5) by change of notation)*

$$\begin{aligned} x_{n+1} &= f_{n+1}(x_1, x_2, \ldots, x_n) \\ &\vdots \\ x_N &= f_N(x_1, x_2, \ldots, x_n) \end{aligned} \tag{6}$$

where the f_i are differentiable functions.

Example

3-5. In the case of the sphere $x^2 + y^2 + z^2 = 1$ in 3-space it has been seen that, around any point, two of the coordinates x, y, z can be used as local coordinates, so that the corresponding projections are the maps defining the local coordinate systems, while the third is given as a differentiable function of them (cf. Example 2-6).

Exercises. **3-1.** Prove the following converse to Theorem 3-1. Let E be Euclidean N-space and let M be a subset such that each point of E has a neighborhood U for which either $U \cap M$ is empty or is the set of points in U satisfying a set of equations that, with suitable numbering of coordinates in E, can be written as

$$\begin{aligned} x_{n+1} &= f_{n+1}(x_1, x_2, \ldots, x_n) \\ &\vdots \\ x_N &= f_N(x_1, x_2, \ldots, x_n) \end{aligned}$$

where the f_i are differentiable. Then prove that M is a submanifold of E.

3-2. Generalize Exercise 3-1 as follows. With the other data as given in the preceding exercise, suppose that each point of E has a neighborhood U such that $U \cap M$ is either empty or is the set of points in U satisfying equations

$$g_i(x_1, x_2, \ldots, x_N) = 0, \qquad i = 1, 2, \ldots, n,$$

where the g_i are differentiable and the Jacobian matrix of the g_i with respect to the x_j is of rank n at each point of U. Prove that M is a submanifold of E.

The two exercises just presented describe situations in which a subset M of Euclidean N-space E turns out to be a submanifold of E. Another such situation will now be discussed. In this case it is rather more difficult to check that condition (2) of Definition 3-1 holds. The idea here is to take the subset of E to be the one-to-one image of a differentiable manifold mapped into E by a differentiable map. It was seen, of course (Example 3-4), that such an image is not necessarily a submanifold in general, but it will now be shown that compactness along with a condition on the rank of the map ensures that it is.

Theorem 3-2. *Let M be a compact differentiable manifold of dimension n and let E be Euclidean N-space. Let $f\colon M \to E$ be a differentiable map that is one-to-one into E and is of rank n at each point of M. Then $f(M)$ is a submanifold of E.*

Proof. Apply Exercise 2-28 to the present situation. Then for any p in M there is a coordinate neighborhood U of p with a local coordinate system ϕ mapping U on an open cell V in Euclidean n-space, where the coordinates are $y_1, y_2, \ldots, y_n,$

and the map $f\phi^{-1}$ is given by equations

$$\begin{aligned}
x_1 &= y_1 \\
&\vdots \\
x_n &= y_n \\
x_{n+1} &= f_{n+1}(y_1, y_2, y_3, \ldots, y_n) \\
&\vdots \\
x_N &= f_N(y_1, y_2, \ldots, y_n)
\end{aligned} \tag{7}$$

where the f_i are differentiable and $x_1, x_2, \ldots, x_N$ denote the coordinates in E. Note that the first n of these equations identify V with an open cell W in the subspace $x_{n+1} = x_{n+2} = \cdots = x_N = 0$ of E, so that the image $f(U)$ appears as the set of points satisfying the equations

$$\begin{aligned}
x_{n+1} &= f_{n+1}(x_1, x_2, \ldots, x_n) \\
&\vdots \\
x_N &= f_N(x_1, x_2, \ldots, x_n)
\end{aligned} \tag{8}$$

with $(x_1, x_2, \ldots, x_n)$ in W and the f_i differentiable.

To complete the proof it must be shown that $f(p)$ has a neighborhood in which the only points of $f(M)$ are those satisfying Eqs. (8).

So let U' be a smaller open neighborhood of p such that $\bar{U}' \subset U$ and write $V' = \phi(U')$ and let W' be the corresponding subset of W. For any positive ϵ the set $W'(\epsilon)$ defined by

$$f_{n+i}(x_1, x_2, \ldots, x_n) - \epsilon < x_{n+i} < f_{n+i}(x_1, x_2, \ldots, x_n) + \epsilon$$
$$(x_1, x_2, \ldots, x_n) \in W'$$

is an open neighborhood of $f(p)$. The idea now is to show that if ϵ is small enough, the only points of $f(M)$ in $W'(\epsilon)$ are those given by the Eqs. (8) with $(x_1, x_2, \ldots, x_n)$ in W'. At first sight it may seem an unnecessary complication to use two neighborhoods U and U' of p, but it is actually a convenience to know that, although we are working with U', the Eqs. (8) remain valid for a bigger set U.

Suppose that the required result is false. Then if a decreasing sequence of values of ϵ were taken, we would get a sequence of points on M, not in U', that would be mapped into the corresponding sets $W'(\epsilon)$. Since M is compact, a convergent subsequence of these points could be picked. Thus, renumbering the subsequence and the corresponding values of ϵ, we would have a sequence $\epsilon_1, \epsilon_2, \ldots$ converging to 0 and points $q_1, q_2, \ldots$ on M converging to q, such that all the q_i are outside U' but $f(q_i)$ is in $W'(\epsilon_i)$ for each i. In fact, all the q_i must be outside U, since no point of $f(U - U')$ can be in $W'(\epsilon)$ for any ϵ, because such points satisfy (8) with $(x_1, x_2, \ldots, x_n)$ not in W'. Thus the limit q is outside U. But since f is continuous, $f(\lim q_i) = \lim f(q_i)$ and, since ϵ_i tends to 0, $\lim f(q_i)$ is in $f(U')$. That is, $f(q)$ is in $f(U')$. But since f is one-to-one, this means that q is in U' and so a contradiction has been reached.

It follows that, for some ϵ, the neighborhood $W'(\epsilon)$ of $f(p)$ satisfies the condition (2) of Definition 3-1. Such a neighborhood can be found for each point of $f(M)$ and so the theorem is completely proved.

3-3. THE EMBEDDING THEOREM

The last section describes differentiable manifolds that are submanifolds of Euclidean space. In fact, no essential restriction is placed on the manifolds in that description, since any differentiable manifold can be represented in this way. The proof of this will be given here for compact manifolds only; it is rather more difficult in the noncompact case.

When a differentiable manifold M is embedded as a submanifold of some Euclidean space E, each coordinate x_i in E appears as a differentiable function on M. Thus the natural way to try to show that a manifold can be embedded in a Euclidean space is to try to construct a set of differentiable functions $f_1, f_2, \ldots, f_N$, and then to consider the map of

M into E defined by mapping p in M on the point with coordinates $(f_1(p), f_2(p), \ldots, f_N(p))$. In the first place, if this map is to be one-to-one, these functions will have to have the property that whenever $p \neq q$, at least one of the f_i has different values at p and q. This property is usually expressed by saying that the functions f must separate the points of M. Obviously, in any one local coordinate neighborhood on M the local coordinates themselves separate points. The idea now is to extend the local coordinates in each local coordinate neighborhood to differentiable functions on the whole manifold in such a way that the extended functions separate points on the whole manifold.

This can be done on a compact manifold M with the help of functions constructed as in Exercise 2-13. So, for any point p of M, construct a neighborhood U of p, contained in a local coordinate neighborhood, and a differentiable function f on M that is positive in U and 0 outside U. Since M is compact, it can be covered by a finite collection $U_1, U_2, \ldots, U_m$ of such neighborhoods, each U_i with its corresponding function f_i. Each U_i is, of course, itself a coordinate neighborhood with coordinate map ϕ_i onto an open cell V_i. If $x_1^{(i)}, x_2^{(i)}, \ldots, x_n^{(i)}$ denote the coordinates of a point in V_i so that each $x_j^{(i)}$ is a differentiable function on U_i, then the functions $y_j^{(i)} = x_j^{(i)} f_i$ are all differentiable functions on M. Consider now the list of functions

$$f_1, y_1^{(1)}, y_2^{(1)}, \ldots, y_n^{(1)}, f_2, y_1^{(2)}, y_2^{(2)}, \ldots, y_n^{(2)}, f_3, \ldots, \\ f_m, y_1^{(m)}, y_2^{(m)}, \ldots, y_n^{(m)}. \quad (9)$$

We will now check that this set of functions separates points on M.

Take two points p and q on M. If p, say, is in U_i but q is not, then f_i is zero at q but not zero at p. Thus f_i separates p and q. On the other hand, suppose that both p and q are in U_i. If $f_i(p) \neq f_i(q)$, then f_i separates p and q. But if $f_i(p) = f_i(q)$, then one of the local coordinates $x_j^{(i)}$ certainly separates p and q and so the corresponding $y_j^{(i)} = f_i x_j^{(i)}$ also does.

Following the idea described at the beginning of this section, the next step should be to use the functions (9) to define a map of M into E. As it stands, it would turn out that the rank of the resulting map is not n at all points. In order that this condition may be satisfied (to enable Theorem 3-2 to be applied), more functions will be added to the list (9). So write $z_j^{(i)}$ for the function equal to $x_j^{(i)}f_i^2$ on U_i and equal to 0 outside U_i. In fact, $z_j^{(i)} = y_j^{(i)}f_i$. Take the expanded list of functions consisting of (9) along with the $z_j^{(i)}$, in all $2mn + n$ functions. Call these functions $F_1, F_2, \ldots, F_N$ with $N = 2mn + n$ and let $F: M \to E$ be the map taking p onto $(F_1(p), F_2(p), \ldots, F_N(p))$. Since the F_i include the functions (9), which already separate the points of M, it follows that F is one-to-one. It will now be checked that F has rank n everywhere. This will be done by showing that, in U_i, at least one of the two $n \times n$ determinants

$$\left|\frac{\partial(y_1^{(i)}, y_2^{(i)}, \ldots, y_n^{(i)})}{\partial(x_1^{(i)}, x_2^{(i)}, \ldots, x_n^{(i)})}\right|, \qquad \left|\frac{\partial(z_1^{(i)}, z_2^{(i)}, \ldots, z_n^{(i)})}{\partial(x_1^{(i)}, x_2^{(i)}, \ldots, x_n^{(i)})}\right|$$

is nonzero at each point.

In fact, remembering that f_i is explicitly given in U_i as the function $\exp(r_i^2 - 1)^{-1}$ where $r_i^2 = \Sigma_{j=1}^n (x_j^{(i)})^2$, we can show, by a straightforward computation, that the first determinant is zero only when $r_i^4 - 4r_i^2 + 1 = 0$ and the second only when $r_i^4 - 6r_i^2 + 1 = 0$, and clearly these equations cannot both be satisfied at any point. Hence, F is of rank n throughout U_i for each i, and so over all of M.

Thus, applying Theorem 3-2, this section can be summed up in the following theorem.

Theorem 3-3. *If M is a compact differentiable manifold, there is a one-to-one mapping $F: M \to E$, where E is a Euclidean space, such that $F(M)$ is a submanifold of E.*

3-4. EMBEDDING A MANIFOLD WITH BOUNDARY

Some minor changes are needed in the foregoing definitions and discussions to cover the case of manifolds with boundaries. For example, if N is a manifold with boundary, condition (2) of Definition 3-7 as stated will apply only to the nonboundary points of N. If p is a point of the boundary of N, the appropriate condition is that it should have a coordinate neighborhood U in M with local coordinate system $\phi: U \to V$ such that

$\phi(N \cap U)$ is the subset of V given by $x_n \geq 0$, $x_{n+1} = x_{n+2} = \cdots = x_m = 0$. The image of the boundary of N in U will be given by the additional condition $x_n = 0$.

If both M and N are manifolds with boundaries, further adjustments are needed to get a definition of N as a submanifold of M, to take care of possible relations between the boundaries.

The proof of the embedding theorem of the last section, with a few minor alterations, shows that if M is a manifold with boundary, there is a one-to-one map $f\colon M \to E$, where E is a Euclidean space, such that $f(M)$ is a submanifold of E. This should be carried out in detail as an exercise.

There is one further adjustment to the embedding of a manifold with boundary in Euclidean space that will be of importance later. It will be convenient to be able to arrange that the boundary lie in a linear subspace. So, suppose M is a compact manifold with boundary and suppose $f\colon M \to E$ is a one-to-one map of M into N-space such that $f(M)$ is a submanifold. Let M_1 be the boundary of M. Suppose that the map f takes p in M onto $(F_1(p), F_2(p), \ldots, F_N(p))$ in E and suppose a differentiable function F_{N+1} can be constructed on M so that it vanishes on M_1 but is positive on $M - M_1$. Then the map f' that maps p in M on the point $(F_1(p), F_2(p), \ldots, F_N(p), F_{N+1}(p))$ is a one-to-one differentiable map of M into Euclidean $(N + 1)$-space E' with the property that $f'(M)$ is a submanifold of E'. Also, the image of f' lies in the set defined by the inequality $x_{N+1} \geq 0$, the image of the boundary M_1 lies in the set $x_{N+1} = 0$, and the image of $M - M_1$ lies in the set $x_{N+1} > 0$.

To construct F_{N+1}, construct a finite covering of M by neighborhoods U_i, each with its corresponding function f_i (as in the proof of Theorem 3-3), f_i being positive on U_i and zero outside U_i. Let $U_1, U_2, \ldots, U_k$ be those coordinate neighborhoods whose corresponding local coordinate systems $\phi_1, \phi_2, \ldots, \phi_k$ map them onto half cells $V_1, V_2, \ldots, V_k$; that is, these are the neighborhoods meeting the boundary of M. If the coordinates on V_i are denoted by $x_1^{(i)}, x_2^{(i)}, \ldots, x_n^{(i)}$ then, say, $x_n^{(i)} \geq 0$ on the image of U_i and the image of $M_1 \cap U_i$ is given by $x_n^{(i)} = 0$. The functions $x_n^{(i)} f_i$, for each i, are differentiable functions on M; thus $\Sigma_{i=1}^k x_n^{(i)} f_i$ is a differentiable function on M. Since all the $x_n^{(i)} f_i$ are nonnegative, it follows that $\Sigma_{i=1}^{k} x_n^{(i)} f_i$ is zero only when all the $x_n^{(i)} f_i$ are zero. For a point in U_i this means that $x_n^{(i)} = 0$ and so the point is in $f(M_1)$. Now define

$$F_{N+1} = \sum_{i=1}^{k} x_n^{(i)} f_i + \sum_{i>k} f_i$$

where the second sum is taken over the f_i corresponding to neighborhoods U_i not meeting M. Here again, all the terms are nonnegative and it is easy to see that F_{N+1} is nonnegative everywhere on M and is zero only on M_1.

So, using F_{N+1} among the mapping functions as described in the foregoing, the following has been proved.

Theorem 3-4. *Let M be a compact differentiable manifold with boundary M_1. Then there is a one-to-one differentiable map f of M into Euclidean $(N + 1)$-space such that $f(M)$ is a submanifold lying in the subset $x_{N+1} \geq 0$, while the intersection of $f(M)$ with the set $x_{N+1} = 0$ is $f(M_1)$.*

Exercises. **3-3.** In the foregoing notation, check that $f(M_1)$ is a submanifold of the Euclidean N-space defined by $x_{N+1} = 0$.

3-4. Adapt the method of proof of Theorem 3-4 to show that if M is a compact differentiable manifold whose boundary is the disjoint union $M_1 \cup M_2$, then there is a one-to-one differentiable map f of M into a Euclidean N-space for some N such that $f(M)$ is a submanifold, lies entirely between the hyperplanes $x_N = 0$ and $x_N = 1$, and has intersections $f(M_1)$ and $f(M_2)$, respectively, with these hyperplanes.

4

Tangent Spaces and Critical Points

4-1. TANGENT LINES

If a curve C in Euclidean N-space is given by parametric equations

$$x_1 = f_1(t),\ x_2 = f_2(t),\ \ldots,\ x_N = f_N(t)$$

where the f_i are differentiable, then the tangent line to C at the point of parameter t_0 is given by the equations

$$\frac{x_1 - f_1(t_0)}{f_1'(t_0)} = \frac{x_2 - f_2(t_0)}{f_2'(t_0)} = \cdots = \frac{x_N - f_N(t_0)}{f_N'(t_0)}.$$

These equations express the geometric idea that the tangent should be the limit of the secant joining the points of parameters t_1 and t_0 as t_1 tends to t_0.

Definition 4-1. If M is a differentiable manifold embedded as a submanifold in a Euclidean space E and if C is a curve contained in M and given by differentiable parametric equations, then the tangent line to C at a point p will be called a *tangent line to M at p.*

Example

4-1. Let M be the sphere $x^2 + y^2 + z^2 = 1$ in 3-space. The planes containing the z axis cut M in circles. The tangents to these circles at (0, 0, 1) are all, by Definition 4-1, tangent lines to M. Note that they all lie in the plane $z = 1$. This remark will be generalized presently.

Lemma 4-1. *Let M be a differentiable manifold of dimension n contained as a submanifold in Euclidean N-space E and let p be a point of M. Then all the tangent lines to M at p lie in an n-dimensional linear subspace T of E.*

Proof. Using Theorem 3-1, assume that the coordinates in E are numbered so that, in a neighborhood of p, M is the set of points satisfying equations of the form

$$x_i = \phi_i(x_1, x_2, \ldots, x_n), \qquad i = n + 1, \ldots, N, \tag{1}$$

with the ϕ_i differentiable. If C has the differentiable parametric equations $x_i = f_i(t)(i = 1, 2, \ldots, N)$ and C is contained in M, then

$$f_i(t) = \phi_i(f_i(t), \ldots, f_n(t)), \qquad i = n + 1, \ldots, N. \tag{2}$$

Assume that C passes through p and that p corresponds to the parameter value t_0. Then Eqs. (2) can be differentiated to give

$$f_i'(t_0) = \sum_{j=1}^{n} \left(\frac{\partial \phi_i}{\partial x_j}\right)_p f'(t_0), \qquad i = n + 1, \ldots, N. \tag{3}$$

But these equations say that every tangent line to M at p lies in the linear space T with the equations

$$x_i - f_i(t_0) = \sum_{j=1}^{n} \left(\frac{\partial \phi_i}{\partial x_j}\right)_p (x_j - f_j(t_0)), \qquad i = n + 1, \ldots, N. \tag{4}$$

The form of these equations shows that T is independent of C and is of dimension n, as required.

Definition 4-2. The linear space T described in Lemma 4-1 is called the *tangent linear space to M at p.*

The following definition introduces a further convenient piece of terminology.

Definition 4-3. If M is a differentiable submanifold of Euclidean N-space E and T is the tangent linear space to M at p, then any hyperplane (linear subspace of dimension $N - 1$) in E containing T will be called *a tangent hyperplane to M at p.*

Note that in case M is of dimension $N - 1$, then T is of dimension $N - 1$ and so there is a unique tangent hyperplane at p, namely, T itself.

Exercises. **4-1.** Find the tangent linear space to the sphere $x^2 + y^2 + z^2 = 1$ in 3-space at a point (x_0, y_0, z_0) on it.

4-2. Let M be a differentiable manifold in Euclidean space and let T be the tangent linear space to M at p. Let L be a line in T through p. Construct a curve C in M with differentiable parametric equations such that L is the tangent to C at p.

Note that Lemma 4-1 simply says that T contains all tangent lines to M at p. This exercise ensures that all lines through p in T are tangent lines to M at p.

4-3. Let M be a differentiable manifold in Euclidean space and let V be a submanifold of M. If T_M and T_V are the tangent linear spaces to M and V, respectively, at a point p of V, prove that $T_V \subset T_M$.

4-2. CRITICAL POINTS

The idea to be introduced here is an extension of the concept of maxima and minima of a function. If a differentiable function f of one variable x has a maximum or minimum for $x = x_0$, then $df/dx = 0$ at x_0. Similarly, if a function f of two variables x, y has a maximum or minimum at (x_0, y_0), then $\partial f/\partial x = \partial f/\partial y = 0$ at this point. What we are really saying here is that the tangent plane to the surface $z = f(x, y)$ (the graph of f in 3-space) is horizontal at the point (x_0, y_0). Clearly, the same condition is also satisfied at a saddle point

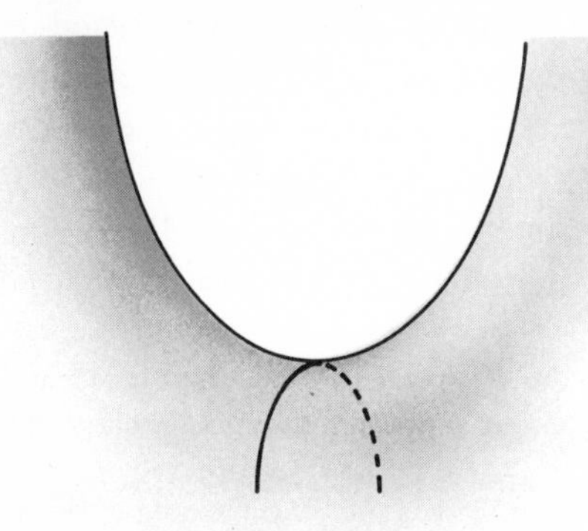

FIGURE 4-1 *Saddle point.*

(cf. Fig. 4-1), a point that behaves like a maximum when approached in one way and like a minimum when approached in another. For functions of more variables there is a wider variety of possibilities described by a similar condition.

Now the surface with equation $z = f(x, y)$ can be thought of as part of a differentiable manifold M on which the local coordinates around $(x_0, y_0, f(x_0, y_0))$ are (x, y); f is then a differentiable function on M with the property that its partial derivatives with respect to local coordinates vanish at $(x_0, y_0, f(x_0, y_0))$. Moreover, this situation can be thought of geometrically as corresponding to an embedding of M in 3-space such that the function f is identified with one of the Euclidean coordinates (namely, z) and such that the horizontal plane $z = f(x_0, y_0)$ is a tangent plane to M at $(x_0, y_0, f(x_0, y_0))$.

The preceding remarks suggest a more general situation, which will now be described.

Let M be an n-dimensional differentiable manifold and let f be a differentiable function on M. Take a local coordinate neighborhood U of a point p of M and let $\phi: U \to V$ be a local coordinate system, V being an open cell in Euclidean n-space. The values at $\phi(p)$ of the first partial derivatives of $f\phi^{-1}$ with respect to the coordinates in V will, of course, depend on the local coordinate system chosen, and will undergo a linear transformation if the coordinate system is changed. However, if these partial derivatives all vanish at $\phi(p)$, the same will be true for any coordinate system.

Definition 4-4. In the notation just introduced, if all the first partial derivatives of $f\phi^{-1}$ with respect to the coordinates in V vanish at $\phi(p)$, p will be called a *critical point of f*.

The remarks just made show that Definition 4-4 is independent of the coordinate system used around p.

Exercises. **4-4.** If M is taken as the plane and f is a differentiable function on M, then any maximum, minimum, or saddle point for f is a critical point.

4-5. Alternatively, if f is as at the beginning of this section but M is taken as the graph of f, namely, the surface $z = f(x, y)$, then any maximum, minimum, or saddle point on M is a critical point of f, regarded as a function on the surface.

A geometric way of looking at critical points, in terms of tangent hyperplanes, will now be described. This corresponds to the introductory remarks to this section on a surface in 3-space.

Let M be a differentiable manifold of dimension n and let f be a differentiable function on M. By Theorem 3-3, no generality is lost by assuming that M is given as a submanifold of a Euclidean space E. Suppose that E is of dimension $N - 1$. Consider now, in N-space E', the set of points $(x_1, x_2, \ldots, x_{N-1}, x_N)$ where $x = (x_1, x_2, \ldots, x_{N-1})$ is a point of M and $x_N = f(x)$. It is easy to see (using Exercise 3-1) that M' is a differentiable submanifold of E', and in fact is diffeomorphic to M under the map taking $(x_1, x_2, \ldots, x_N)$ onto $(x_1, x_2, \ldots, x_{N-1})$. The point is that M' is now a submanifold of E' with the property that the given differentiable function f coincides with the Euclidean coordinate x_N.

Now take a point p' on M' with coordinates $(x_1^0, x_2^0, \ldots, x_N^0)$ and let p be the corresponding point $(x_1^0, x_2^0, \ldots, x_{N-1}^0)$ of M; p has a neighborhood U in E such that the points of $M \cap U$ are the points satisfying equations of the form (with suitable numbering of the coordinates)

$$x_i = f_i(x_1, x_2, \ldots, x_n), \qquad i = n + 1, \ldots, N - 1, \quad (5)$$

where the f_i are differentiable (cf. Theorem 3-1). And so, in a neighborhood of p', M' is the set of points satisfying the Eqs.

(5) along with

$$x_N = f(x_1, x_2, \ldots, x_n).$$

Here $x_1, x_2, \ldots, x_n$ can be used as local coordinates on M or M' around p or p', respectively. The equations of the tangent linear space to M' at p' are then

$$\begin{aligned} x_i - x_i^0 &= \sum_{j=1}^{n} \left(\frac{\partial f_i}{\partial x_j}\right)_p (x_j - x_j^0), \qquad i = n+1, \ldots, N-1, \\ x_N - x_N^0 &= \sum_{j=1}^{n} \left(\frac{\partial f}{\partial x_j}\right)_p (x_j - x_j^0). \end{aligned} \tag{6}$$

If now p is a critical point of f on M or, what is the same thing, if p' is a critical point of f on M', the last equation of (6) becomes

$$x_N = x_N^0 = f(p). \tag{7}$$

That is, (7) is the equation of a tangent hyperplane to M' at p'. Conversely, if (7) is a tangent hyperplane to M' at p', then all the coefficients on the right of the last equation of (6) are zero, and so p' is a critical point of f.

This can all be summed up as follows (using the notation in the foregoing, but dropping the primes).

Lemma 4-2. *If M is a differentiable manifold and f a differentiable function on it, then M can be embedded as a submanifold of a Euclidean N-space such that the value of f at a point is the value of x_N at that point, and f has a critical point at p if and only if $x_N = f(p)$ is a tangent hyperplane to M at p.*

Of course, it may happen that the function f is not given initially. We may simply be given a submanifold M of a Euclidean N-space. Then if f is defined as x_N, the conclusion of Lemma 4-2 still holds, namely, that critical points of x_N are points at which there is a tangent hyperplane of the form $x_N =$ constant.

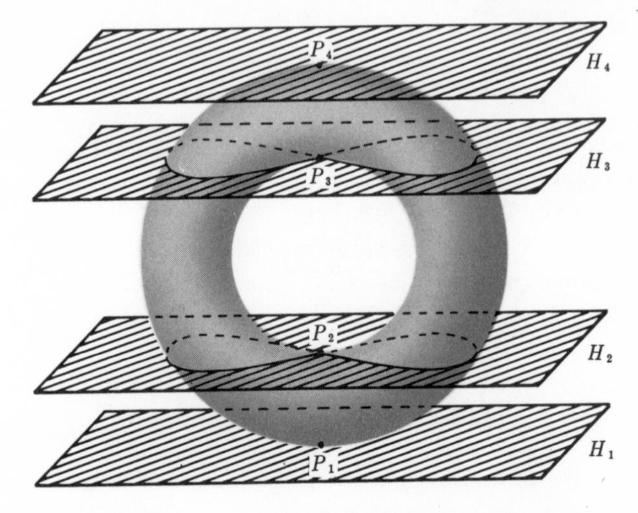

FIGURE 4-2 *Four critical points of z on the torus.*

Example

4-2. Let M be a torus embedded as in Section 2-1 as a submanifold of Euclidean 3-space. It is easy to see from Fig. 4-2 that there are just four horizontal planes (that is, planes of the form $z =$ constant) H_1, H_2, H_3, H_4 that are tangent planes of M at P_1, P_2, P_3, P_4, respectively, corresponding to the fact that the function z on M has critical points at P_1, P_2, P_3, P_4.

Exercise. **4-6.** Verify the last remark (that z has critical points at the P_i) by expressing z in terms of the local coordinates around these points.

Note that, in the preceding example, the critical points of z are points around which x and y are local coordinates, and z could not be used as one of the local coordinates around any of these points. This is exactly what we should expect. For in any neighborhood where z is one of the local coordinates, $\partial z/\partial z = 1$, and so the condition of Definition 4-4 for a critical point could not be satisfied.

On the other hand, it will be noticed that P_1, P_2, P_3, P_4 are the only points around which z cannot be used as one of the local coordinates. In addition, if C is the section of the torus by a horizontal plane $z = c$ other than H_1, H_2, H_3, H_4, then in a local coordinate neighborhood of a point on C, C is the set satisfying $z - c = 0$, with z, or $z - c$, as one of the local coordinates. Hence, C is a submanifold of the torus.

The properties just described will now be formulated in general for any differentiable function on a differentiable manifold.

Lemma 4-3. *Let M be a differentiable manifold of dimension n, f a differentiable function on M. Let p be a point of M that is not a critical point of f. Then there is a local coordinate system $\phi\colon U \to V$, where V is an open n-cell in which the coordinates are $x_1, x_2, \ldots, x_n$, such that $f\phi^{-1} = x_1$. If p is a critical point of f, then no local coordinate system around p has this property.*

Proof. Suppose p is not a critical point of f and let $\phi'\colon U' \to V'$ be a local coordinate system in the neighborhood U' of p. Let coordinates in V' be denoted by $y_1, y_2, \ldots, y_n$ and write $f\phi'^{-1} = g(y_1, y_2, \ldots, y_n)$. Then, since p is not a critical point of f, one of the partial derivatives, say $\partial g/\partial y_1$, is not zero at $\phi(p)$. Thus the functions $x_1, x_2, \ldots, x_n$ defined by

$$x_1 = g(y_1, y_2, \ldots, y_n)$$

$$x_i = y_i, \qquad i = 2, 3, \ldots, n$$

have a nonzero Jacobian determinant with respect to $y_1, y_2, \ldots, y_n$ at $\phi'(p)$. It follows (cf. Definition 2-7) that there is a local coordinate system $\phi\colon U \to V$ such that $x_1, x_2, \ldots, x_n$ are the coordinates in V and $f\phi^{-1} = x_1$, as required.

Conversely, if there is a coordinate system with this property, then $\partial/\partial x_1\,(f\phi^{-1}) = 1$ at $\phi(p)$ and so p is not a critical point of f.

Corollary. *For a constant c let M_c be the set of points, in the notation of Lemma* 4-3, *at which f has the value c. Then if M_c contains no critical point of f, it is a submanifold of M. Its dimension is* $\dim M - 1$.

Proof. Take p on M_c. Since p is not a critical point of f, a local coordinate system $\phi\colon U \to V$ can be taken around p, as in Lemma 4-3, such that $f\phi^{-1} = x_1$. Thus the set $\phi(M_c \cap U)$

is the set of points in V such that $x_1 - c = 0$. And since $x_1 - c$ can be taken as one of the coordinates in V, condition (2) of Definition 3-1 is satisfied, showing that M_c is a submanifold of dimension $\dim M - 1$.

Expressed in terms of an embedding of M in Euclidean space, as in Lemma 4-2, the last result means that sections of M by hyperplanes $x_N =$ constant not containing critical points of x_N are all submanifolds of M of dimension $\dim M - 1$. The hyperplanes that fail to satisfy this condition are tangent hyperplanes to M.

4-3. NONDEGENERATE CRITICAL POINTS

In the Example 4-2 of the torus in 3-space, the function z, expressed in terms of x and y on the surface around the critical points P_i, has a rather special form. If the torus M is thought of as the surface traced by the circle of center (2, 0) and radius 1 in the (x, y) plane as this plane is rotated about the y axis, then its equation is

$$(x^2 + y^2 + z^2 + 3)^2 = 16(x^2 + z^2).$$

Write this equation as a quadratic in z^2 and solve for z^2. The result is

$$z^2 = 5 - x^2 - y^2 \pm 4(1 - y^2)^{1/2}. \tag{8}$$

According as the plus or minus sign is taken, this gives two values for z^2 and thus four values for z. For small values of x and y these correspond to the local equations of M in neighborhoods of the points P_i. For example, to get the local equation around P_3, take the minus sign in (8) and expand the right-hand side in a power series. This expansion is

$$z^2 = 1 + y^2 - x^2 + 3y^4 + \cdots .$$

And so, taking the positive square root and expanding in a power series, we get

$$z = 1 + \tfrac{1}{2}(y^2 - x^2) + \cdots ,$$

where the remaining terms are all of degree greater than two.

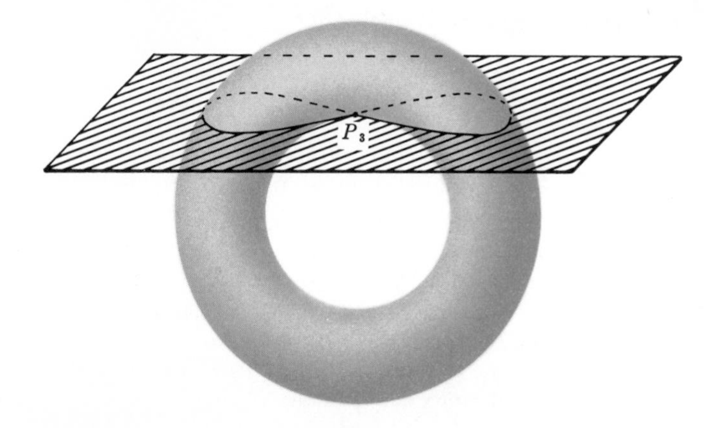

FIGURE 4-3 *A critical level for the function z on the torus. The first approximation at P_3 is $z = 1 + \frac{1}{2}(y^2 - x^2)$.*

Thus a first approximation for the equation of M around P_3 is obtained by ignoring the higher powers of x and y. This approximation is $z = 1 + \frac{1}{2}(y^2 - x^2)$ and the section by $z = 1$ is $y = \pm x$, the pair of tangents at the crossover of the figure eight in which this plane cuts the torus (see Fig. 4-3). The important thing to notice is that the first approximation to the equation of M at P is given by writing $z - 1$ as a quadratic in x and y which is nondegenerate; that is, whose determinant is not zero. This determinant is the value of

$$\begin{vmatrix} \dfrac{\partial^2 z}{\partial x^2} & \dfrac{\partial^2 z}{\partial x\,\partial y} \\ \dfrac{\partial^2 z}{\partial x\,\partial y} & \dfrac{\partial^2 z}{\partial y^2} \end{vmatrix}$$

at P_3.

These considerations motivate the following definition.

Definition 4-5. Let f be a differentiable function on a differentiable manifold M of dimension n and let p be a critical point of f. Let U be a local coordinate neighborhood of p and $\phi\colon U \to V$ be a local coordinate system. Write $g = f\phi^{-1}$, a function of the n coordinates $x_1, x_2, \ldots, x_n$ in the cell V.

If the determinant

$$\left|\frac{\partial^2 g}{\partial x_i\,\partial x_j}\right|, \qquad i, j = 1, 2, \ldots, n,$$

is not zero at $\phi(p)$, the critical point p will be called *nondegenerate*. Note that the rank of the matrix $(\partial^2 g/\partial x_i\,\partial x_j)$ at $\phi(p)$ is thus the dimension of M.

Exercises. **4-7.** Verify that Definition 4-5 is independent of the local coordinate system used around p.

4-8. Prove that, in Example 4-2, P_1, P_2, P_3, P_4 are nondegenerate critical points of the function z on the torus.

4-9. Take the torus M, as in Example 4-2, and consider y as a function on M. Check that the planes $y = \pm 1$ are tangent planes to M, each touching M along a circle, and so show that all points of these circles are critical points for y. But show that these critical points are degenerate (that is, not nondegenerate).

4-10. It will be noted in Exercise 4-9 that the degenerate critical points are not isolated. However, this state of affairs does not necessarily go with degeneracy. For example, consider the function $x^3 + y^3$ on the plane. Show that $(0, 0)$ is an isolated critical point, but degenerate.

4-11. Let f be a differentiable function of n variables $x_1, x_2, \ldots, x_n$ vanishing when $x_1 = x_2 = \cdots = x_n = 0$. Note that

$$f(x_1, x_2, \ldots, x_n) = \int_0^1 \frac{d}{dt} f(tx_1, tx_2, \ldots, tx_n)\, dt$$

and hence show that $f = \sum x_i f_i$, where the f_i are differentiable functions such that $f_i(0, 0, \ldots, 0)$ is the value of $\partial f/\partial x_i$ at $x_1 = x_2 = \cdots = x_n = 0$.

If in addition all the $\partial f/\partial x_i$ vanish when all the x_j are zero, show that $f = \sum x_i x_j f_{ij}$, where the f_{ij} are differentiable functions such that $f_{ij}(0, 0, \ldots, 0)$ is the value of $\partial^2 f/\partial x_i\,\partial x_j$ at $x_1 = x_2 = \cdots = x_n = 0$.

This exercise means that if f is a differentiable function on a manifold M with a critical point at p, and $\phi\colon U \to V$ is a local coordinate system in a neighborhood of p, $\phi(p)$ being the point $(0, 0, \ldots, 0)$, then $g = f\phi^{-1}$ is a quadratic in the x_i, the coordinates in V, with coefficients that coincide with the $\partial^2 g/\partial x_i\,\partial x_j$ at $\phi(p)$. It would clearly be a simplification if local coordinates could be constructed in terms of which f would have an expression as a quadratic with constant coefficients. The next exercise will show that this can be done.

4-12. As in Exercise 4-11 let f be a differentiable function of n variables $x_1, x_2, \ldots, x_n$ such that f and all the $\partial f/\partial x_i$ are zero when $x_1 = x_2 = \cdots = x_n = 0$, and write $f = \sum x_i x_j f_{ij}(x)$. Show that there is a transformation of variables of the form

$$y_i = \sum h_{ij}(x) x_j$$

where the h_{ij} are differentiable and the determinant $|h_{ij}(0)| \neq 0$, such that

$$f = \sum c_i y_i^2$$

and the c_i are all 1, -1, or 0.

[Start by diagonalizing $\sum f_{ij} x_i x_j$ just as a quadratic function with constant coefficients. That is, arrange the x_i so that $f_{11}(0) \neq 0$ and remove the product terms $x_1 x_i$ by writing

$$y_1 = \sqrt{|f_{11}|}\left(x_1 + \frac{f_{12}}{f_{11}} x_2 + \cdots + \frac{f_{1n}}{f_{11}} x_n\right).$$

Then proceed by induction. Check that the coefficients of the resulting transformation $y_i = \sum h_{ij}(x) x_j$ satisfy the required conditions.]

The following theorem is an immediate consequence of the results of the preceding exercises.

Theorem 4-1. *Let f be a differentiable function on a differentiable manifold M and let p be a nondegenerate critical point of f on M. Then there is a local coordinate system $\phi\colon U \to V$ on a neighborhood of p such that $f\phi^{-1}$, expressed in terms of the coordinates in V, is $\Sigma c_i y_i^2$ where each c_i is either* 1 *or* -1. *Here $\phi(p)$ is the point $y_1 = y_2 = \cdots = y_n = 0$.*

Proof. Take any local coordinate system $\chi\colon U' \to V'$ about p and suppose that the coordinates in the cell V' are denoted by $x_1, x_2, \ldots, x_n$, $\chi(p)$ being the point $(0, 0, \ldots, 0)$, and that $f\chi^{-1}$ is the function $g(x_1, x_2, \ldots, x_n)$. Make the change of variables from the x_j to the y_i as in Exercise 4-12. The Jacobian determinant of the y_i with respect to the x_j is the determinant $|h_{ij}(x)|$, which is not zero at $\chi(p)$, and so by Definition 2-7 there is a local coordinate system $\phi\colon U \to V$ with the y_i as coordinates in V such that $f\phi^{-1} = \Sigma c_i y_i^2$. Each c_i will be 1.

-1, or 0, but the value 0 is excluded by the assumption of nondegeneracy of p.

The index of inertia of the form $\Sigma c_i y_i^2$ appearing in Theorem 4-1 (that is, the number of positive c_i minus the number of negative c_i) is an invariant under linear transformation of the variables. It follows that it depends only on the function f and the point p. Since, for a nondegenerate critical point, the rank of the quadratic $\Sigma c_i y_i^2$ is the dimension of M, this means that the number r of negative c_i depends only on p and the function f, and not on the particular local coordinates used.

Definition 4-6. The number r just introduced associated with the nondegenerate critical point p of f will be called the *type number of the critical point.*

4-4. A STRONGER EMBEDDING THEOREM

In Exercises 4-8 and 4-9 two functions, y and z, were considered on the torus, one with degenerate critical points, the other with nondegenerate critical points. A more convenient arrangement here will be to interchange the y and z axes in Exercise 4-9. Thus we have two embeddings of the torus in 3-space, namely, as the surface

$$(x^2 + y^2 + z^2 + 3)^2 = 16(x^2 + z^2)$$

and as the surface $(x^2 + y^2 + z^2 + 3)^2 = 16(x^2 + y^2)$. In each case the Euclidean coordinate z is a differentiable function on the embedded torus, in the first case with four nondegenerate critical points (see Fig. 4-2) and in the second with infinitely many degenerate critical points (see Fig. 4-4). Thinking of the nondegenerate critical points as the simpler kind, the first embedding is thus to be regarded as better than the second. In fact, a bit of computation shows that the horizontal embedding of Fig. 4-4 is the only bad embedding of the torus. If the surface is rotated just a little bit out of this position, it will turn out that z has four nondegenerate critical points.

The situation just described is a special case of the following general theorem.

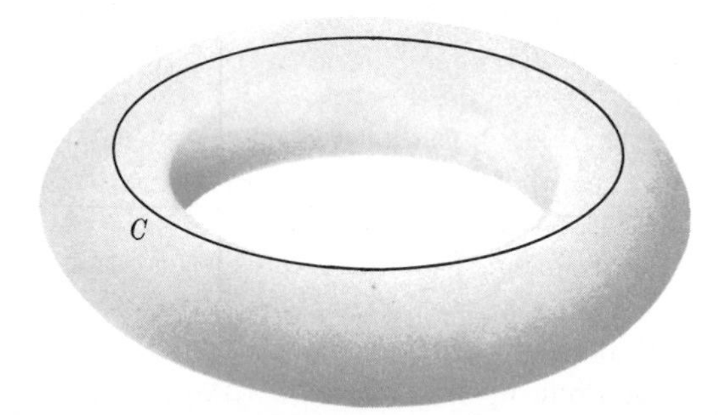

FIGURE 4-4 *Circle C of degenerate critical points for the function z on the horizontally embedded torus.*

Theorem 4-2. *Let M be a compact differentiable manifold with boundary, the boundary consisting of a disjoint union $M_1 \cup M_2$. Then there is a one-to-one differentiable map f of M into a Euclidean N-space, for some N, with the following properties.*

(1) $f(M)$ is a submanifold of N-space.

(2) $f(M)$ lies entirely in the set $0 \leq x_N \leq 1$, and the intersections of $f(M)$ with the hyperplanes $x_N = 0$ and $x_N = 1$ are $f(M_1)$ and $f(M_2)$, respectively.

(3) The function x_N on $f(M)$ has only a finite number of critical points on $f(M)$, none lying on $x_N = 0$ or $x_N = 1$, and they are all nondegenerate. It can also be arranged that no two critical points correspond to the same value of x_N.

Part of condition (3) can be stated by saying that, of the hyperplanes $x_N = c$, only a finite number are tangent hyperplanes to $f(M)$, each such tangent hyperplane corresponding to just one point of contact.

There is another way of formulating the theorem, of course. The function x_N on $f(M)$ composed with the map f is a differentiable function ϕ on M. Thus the theorem says that there is a differentiable function ϕ on M taking the values 0 and 1 on M_1 and M_2, respectively, satisfying $0 < \phi(p) < 1$ at all nonboundary points of M, and having only a finite number of critical points on M, all corresponding to different values of ϕ, all nondegenerate, and none lying on M_1 or M_2.

Note also that if M should be a compact differentiable manifold without boundary, then a similar result holds, but with no mention of boundaries.

The proof of Theorem 4-2 is rather difficult and will not be given here. But the formulation given suggests a possible approach to the proof. For it has already been seen that, given the manifold M with boundary $M_1 \cup M_2$, there is a one-to-one map f into a Euclidean space satisfying conditions (1) and (2). The idea, then, is to show that this map f can be adjusted so that condition (3) is also satisfied.

This can be done by first showing that f can be approximated by a map f' such that $f'(M)$ is part of an algebraic variety, that is, the set of zeros of a finite set of polynomial equations in the Euclidean coordinates $x_1, x_2, \ldots, x_N$. Note that in the examples given of embedding of the torus in 3-space, this stage has already been reached. That is, in each case the torus is given as an algebraic surface.

The second step of the proof is to show that the coordinates in N-space can be adjusted so that x_N satisfies condition (3) relative to $f'(M)$. The idea here is to consider the function $\Sigma u_i x_i$ where the u_i are real numbers. It turns out that the condition that this function should have infinitely many critical points or degenerate critical points on M is that the u_i should satisfy certain polynomial equations. So choose a set of values of the u_i not satisfying these equations, and then make a change of coordinates so that $\Sigma u_i x_i$ becomes the new Nth coordinate. Condition (3) will now be satisfied.

5

Critical and Noncritical Levels

5-1. DEFINITIONS AND EXAMPLES

Suppose that M is a compact differentiable manifold with boundary $M_0 \cup M_1$, embedded in Euclidean N-space so that it lies entirely between the hyperplanes $x_N = 0$ and $x_N = 1$, its intersections with these hyperplanes being M_0 and M_1, respectively. A careful study is now to be made of the relations between the various sections of M by hyperplanes $x_N =$ constant. It would come to the same thing to study the level sets of a differentiable function on M without reference to a Euclidean space, but the Euclidean embedding automatically provides a notion of perpendicularity that will be needed presently.

Definition 5-1. If the section M_c of M by the hyperplane $x_N = c$ contains a critical point of x_N, M_c will be called *a critical level of* x_N. Otherwise it will be called a *noncritical level of* x_N.

For instance, in Example 4-2 the function z on the torus has four critical levels, namely, the sections by H_1, H_2, H_3, H_4. Now it will be noticed in that example that, if M_c is a noncritical level of z, it is surrounded by neighboring noncritical levels, all of which are homeomorphic to each other. For example (see Fig. 5-1), between H_1 and H_2 all the noncritical levels are circles. On the other hand, as we cross a critical level, a change takes place. The noncritical levels immediately below H_2 are different from those immediately above. The main object of this chapter is to show that the remarks just made are valid in general, and further, to examine exactly what happens as we make the transition from one side of a critical level to the other.

This study will be carried out with the help of the family of orthogonal trajectories of the sections M_c, that is, a family of curves in M with differentiable parametric equations, cutting the M_c at right angles (except at critical points). The idea can be illustrated by the following rather trivial example.

Let M be a 2-sphere in Euclidean 3-space. The function z has two critical points, namely, the north and south poles. The noncritical levels of z are the circles of latitude. The orthogonal trajectories of these are the circles of longitude. One can think of one noncritical level as being mapped on another by sliding the points of one along the circles of longitude till they reach the other. In addition, if M_c is a noncritical level, a neighborhood of it can be expressed as a union

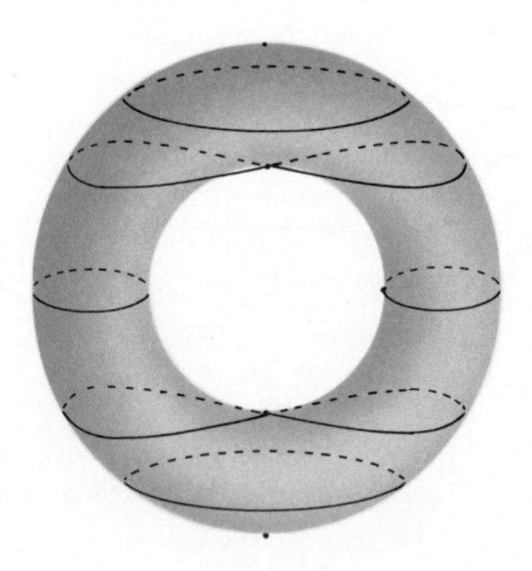

FIGURE 5-1 *Critical and noncritical levels for the function z on the torus.*

of arcs of the circles of longitude so that it has the form $M_c \times I$, where I is a line interval. It will turn out that a similar situation holds in general. Then, of course, the behavior of the orthogonal trajectories near a critical point will have to be studied. As can already be seen from the example just given, it will no longer be true that a critical point lies on just one trajectory.

In the first place, the family of orthogonal trajectories to the M_c must be constructed in the general case. This will be done by setting up differential equations for them and then appealing to the appropriate existence theorem from the theory of differential equations.

So start off with a neighborhood U in Euclidean N-space in which M is the set of points satisfying equations of the form

$$\begin{aligned} x_{n+1} &= f_{n+1}(x_1, x_2, \ldots, x_n) \\ &\vdots \\ x_N &= f_N(x_1, x_2, \ldots, x_n) \end{aligned}$$

where the f_i are differentiable functions. If $U \cap M$ contains no critical point of x_N, Lemma 4-3 implies that x_N itself can be taken as one of the local coordinates in $U \cap M$. This means that, with suitable renumbering of the coordinates, $U \cap M$ will be the set of points satisfying equations of the form

$$x_i = f_i(x_1, x_2, \ldots, x_{n-1}, x_N), \qquad i = n, n+1, \ldots, N-1. \tag{1}$$

The set $U \cap M_c$, if nonempty, will then be given by Eqs. (1) with the additional equation $x_N = c$.

Let p be a point of $U \cap M$. Use the notation $x_i(p)$ for the value of the coordinate x_i at p. Thus, for the level M_c through p, $c = x_N(p)$. Let T_p be the tangent linear space of M at p, T'_p that of M_c at p. Then T_p and T'_p are of dimensions n and $n - 1$, respectively, and $T'_p \subset T_p$ (cf. Exercise 4-3). Since p is not a critical point for x_N (for there is no critical point in U), T_p does not lie in the hyperplane $x_N = c = x_N(p)$. On the other hand, T'_p clearly does, and so a direction in T_p orthogonal to T'_p is not in the hyperplane $x_N = x_N(p)$. Thus a curve in M through p and with tangent line orthogonal to T'_p should have parametric equations of the form

$$x_i = \phi_i(x_N), \qquad i = 1, 2, \ldots, N-1$$

and its tangent line will have direction ratios

$$\phi_1'(x_N), \qquad \phi_2'(x_N), \ldots, \phi_{N-1}'(x_N), \qquad 1, \tag{2}$$

the derivatives all being evaluated at p.

The equations of T_p are

$$x_i - x_i(p) = \sum_{j=1}^{n-1} \left(\frac{\partial f_i}{\partial x_j}\right)_p (x_j - x_j(p)) + \left(\frac{\partial f_i}{\partial x_N}\right)_p (x_N - x_N(p)) \tag{3}$$

where the subscript p on the right-hand side denotes evaluation at p. The equations of T_p' are

$$x_i - x_i(p) = \sum_{j=1}^{n-1} \left(\frac{\partial f_i}{\partial x_j}\right)_p (x_j - x_j(p)). \tag{4}$$

Since the direction (2) is to be orthogonal to T_p', it must be orthogonal to a set of $n-1$ linearly independent directions in T_p'. Such a set of directions will be a set of linearly independent solutions of the Eqs. (4), regarded as linear equations in the $x_i - x_i(p)$. The following $n-1$ sets of direction ratios form such a set of solutions:

$$\begin{array}{l}
1, 0, 0, \ldots, 0, \left(\frac{\partial f_n}{\partial x_1}\right)_p, \left(\frac{\partial f_{n+1}}{\partial x_1}\right)_p, \ldots, \left(\frac{\partial f_{N-1}}{\partial x_1}\right)_p, 0 \\
0, 1, 0, \ldots, 0, \left(\frac{\partial f_n}{\partial x_2}\right)_p, \left(\frac{\partial f_{n+1}}{\partial x_2}\right)_p, \ldots, \left(\frac{\partial f_{N-1}}{\partial x_2}\right)_p, 0 \\
\quad \cdot \qquad\qquad \cdot \qquad\qquad \cdot \\
\quad \cdot \qquad\qquad \cdot \qquad\qquad \cdot \\
\quad \cdot \qquad\qquad \cdot \qquad\qquad \cdot \\
0, 0, 0, \ldots, 1, \left(\frac{\partial f_n}{\partial x_{n-1}}\right)_p, \left(\frac{\partial f_{n+1}}{\partial x_{n-1}}\right)_p, \ldots, \left(\frac{\partial f_{N-1}}{\partial x_{n-1}}\right)_p, 0.
\end{array} \tag{5}$$

Thus the direction (2) must be orthogonal to these, and must also satisfy the Eqs. (3) when substituted for the $x_i - x_i(p)$ (thus ensuring that it lies in T_p).

***Exercise.* 5-1.** Prove that the resulting set of linear equations satisfied by (2) has a unique solution, expressing the $d\phi_i/dx_N(i = 1, 2, \ldots,$

$N - 1$) as differentiable functions on M with the critical points of x_N removed.

Now there is a theorem in differential equation theory (cf. [5]) that guarantees that a set of differential equations such as has just been obtained, expressing the $d\phi_i/dx_N$ as differentiable functions, has a set of solutions consisting of a family of curves F on M, with the critical points of x_N removed, with the property that exactly one curve of F passes through each point. The method of derivation of the differential equations shows that the family F is the required family of orthogonal trajectories to the M_c. The same theorem on differential equations gives further information, namely, that the equations of a member of F in a neighborhood on M not containing a critical point of x_N can be written as

$$x_i = \phi_i(x_1^0, x_2^0, \ldots, x_{n-1}^0, x_N), \tag{6}$$

where x_N is used as the parameter on the curve, and $x_1^0, x_2^0, \ldots, x_{n-1}^0$ are the local coordinates of the point in which the curve in question meets some fixed M_c, and the ϕ_i are differentiable in all the variables.

Exercises. **5-2.** Prove that (6) can be used to introduce a new coordinate system in a neighborhood of a noncritical point of x_N, specifying a point by its parameter x_N on the appropriate curve of F, along with the coordinates of the point in which that curve meets a fixed M_c.

Hence, show that if M_c is a noncritical level of x_N, it has a neighborhood on M that is diffeomorphic to $M_c \times I$, where I is an open interval of values of x_N. Note that a similar result will also hold for a critical level if a neighborhood of the critical point is removed.

5-3. Deduce from the last exercise that if M_{c_1} and M_{c_2} are two consecutive critical sections (that is, such that all the levels M_c with $c_1 < c < c_2$ are noncritical), then the part of M between M_{c_1} and M_{c_2} is diffeomorphic to $M_c \times I$ where $c_1 < c < c_2$.

5-4. Also deduce from Exercise 5-2 that if two noncritical levels have no critical level between them, they are diffeomorphic.

5-2. A NEIGHBORHOOD OF A CRITICAL LEVEL; AN EXAMPLE

The results of the last section show that as c increases from 0 to 1, the section M_c of M by $x_N = c$ remains the same (diffeomorphic manifolds being regarded as the same) so long as we stay between two critical levels. However, M_c changes when we cross a critical level, and the next step is to investigate the nature of this change.

So let M_c be a critical level for x_N and let P be the critical point on it. As already noted, the orthogonal trajectories F around any point of M_c except P behave in essentially the same way as around a noncritical level. The important thing, then, is to examine their behavior around P.

To introduce the discussion, consider the function z on the torus M, embedded in Euclidean 3-space as in Example 4-2, and in particular fix attention on a neighborhood of the critical point P_3. It has already been remarked (Section 4-3) that a first approximation to the torus near this point is the surface $z = 1 + \frac{1}{2}(y^2 - x^2)$. This is made more precise by Theorem 4-1. An alternative way of stating that result is: There is a diffeomorphism ϕ of a neighborhood of P_3 on the torus onto a neighborhood of the point (0, 0, 0) of the surface $z = y^2 - x^2$ (the factor $\frac{1}{2}$ is removed by a suitable change of scale). Of course, ϕ will not necessarily carry orthogonal trajectories to the horizontal sections of the torus into orthogonal trajectories to the horizontal sections of $z = y^2 - x^2$. On the other hand, it is convenient to study the latter, so that, in effect, we will have a standard description of a neighborhood of any saddle point. To allow this, it will be shown that the orthogonal trajectories F to the horizontal sections of the torus can be adjusted in a neighborhood of P_3 to give a new family F' with similar properties, but such that ϕ carries F' in a neighborhood of P_3 into the family F_0 of orthogonal trajectories to the horizontal sections of $z = y^2 - x^2$.

The device to be used is quite simple. At each point p near P_3 on the torus (but not equal to P_3) there are two sets of direction ratios assigned, those of the tangent to the curve of the family F through p and those of the tangent to the curve of the family $\phi^{-1}(F_0)$ through p. Denote these sets of ratios by $l_1, l_2, 1$ and $\lambda_1, \lambda_2, 1$, respectively. Now assign to p the direction given by

$$g\lambda_1 + (1 - g)l_1, g\lambda_2 + (1 - g)l_2, 1, \tag{7}$$

where g is a differentiable function on the torus equal to zero outside a neighborhood of P_3 and to 1 inside a smaller neighborhood. Then construct a new family of curves F' tangent to this direction at p. Note that as we approach P_2 the direction (7) gradually changes from $l_1, l_2, 1$ to $\lambda_1, \lambda_2, 1$. Thus the family F' will be equal to F outside a neighborhood of P_3 but will coincide with $\phi^{-1}(F_0)$ in a smaller neighborhood.

As pointed out earlier, this means that a neighborhood of H_2 on the torus can be studied by using the adjusted orthogonal trajectories F', and this in turn means that a neighborhood of P_3 can be studied by looking at the origin on the surface $z = y^2 - x^2$, using the curves F_0 on that surface.

5-3. NEIGHBORHOOD OF A CRITICAL LEVEL; GENERAL DISCUSSION

The ideas illustrated in the last section by a two-dimensional example will now be formulated in general. That is, a neighborhood of a nondegenerate critical point of a function on a given manifold will be compared with a neighborhood of some kind of standard critical point (corresponding to the origin on the surface $z = y^2 - x^2$).

So let M be a differential manifold of dimension n, let f be a differentiable function on M, and let p be a nondegenerate critical point of f. By adding a suitable constant it can be assumed that $f(p) = 0$. Use Theorem 4-1 to set up a local coordinate system $\psi: U \to V$ in a neighborhood U of p such that

$$f\psi^{-1} = \sum_{i=1}^{n} c_i y_i^2 \tag{8}$$

where the y_i are the Euclidean coordinates in V and each c_i is ± 1. The y_i are all zero at $\psi(p)$.

At the same time, in a Euclidean space of dimension $n + 1$, in which the coordinates are denoted by $(y_1, y_2, \ldots, y_n, z)$, consider the hypersurface Q with the equation

$$z = \Sigma c_i y_i^2. \tag{9}$$

V can be identified with a neighborhood of the origin in the hyperplane $z = 0$. Write V' for the neighborhood on Q that projects on V. Then define the map $\phi: U \to V'$ by mapping q on $(y_1, y_2, \ldots, y_n, z)$ where $\psi(q)$ is the point $(y_1, y_2, \ldots, y_n)$ and z is given by (9). Note now that the value of $f\phi^{-1}$ at

$(y_1, y_2, \ldots, y_n, z)$ on Q is the same as that of $f\psi^{-1}$ at $(y_1, y_2, \ldots, y_n)$, namely, $\Sigma c_i y_i^2$, and by (9) this is z.

Thus $\phi: U \to V'$ is a map into an open set on Q such that the value of $f\phi^{-1}$ is the value of z. In particular, this means that the subsets f = constant in U are mapped into the sections of Q by the hyperplanes z = constant.

Next assume (cf. Lemma 6-2) that M is embedded as a submanifold in some Euclidean N-space, such that the function f on M is x_N, and construct, as in Section 5-1, the orthogonal trajectories F to the sections of M by the hyperplanes x_N = constant. The F are now to be adjusted in a neighborhood of p as illustrated in Section 5-2.

So let F_0 be the family of orthogonal trajectories on the quadratic hypersurface Q to the family of sections by the hyperplanes $z = c$. F_0 can be defined at all points except the origin. $\phi^{-1}(F_0)$ will then denote the image in U of the family F_0 in V. Thus through any point q in U other than p there are two curves, one of the family F and one of the family $\phi^{-1}(F_0)$. Write the direction ratios of their tangent lines as

$$l_1(q), l_2(q), \ldots, l_{N-1}(q), 1, \tag{10}$$

$$\lambda_1(q), \lambda_2(q), \ldots, \lambda_{N-1}(q), 1, \tag{11}$$

respectively. Note here that it is admissible to take the last ratio as 1 in each case, since neither of these directions is horizontal.

Next let U_1 be a second neighborhood of p such that $\bar{U}_1 \subset U$. For the sake of simplicity U_1 may as well be a cell. Define (cf. Exercise 2-14) a differentiable function g on M such that $g = 1$ on $\bar{U}_1$ and $g = 0$ outside U. Then to each point q of U assign the set of direction ratios

$$g(q)\lambda_1(q) + (1 - g(q))l_1(q), \ldots, g(q)\lambda_{N-1}(q) + (1 - g(q))l_{N-1}(q), 1. \tag{12}$$

Note that this assignment extends naturally outside U simply by assigning $l_1, l_2, \ldots, l_{N-1}, 1$ to a point outside U. Then, to find a family of curves F' whose tangent directions at each point are given by (12), solve the differential equations

$$\left(\frac{dx_i}{dx_N}\right)_y = g(q)\lambda_i(q) + (1 - g(q))l_i(q), \qquad i = 1, 2, \ldots, N - 1. \tag{13}$$

Since the right-hand sides are differentiable functions at all points of M except critical points of x_N, the existence theorem previously quoted (cf. [5]) shows that there is a family F' of differentiable curves on M, one through each point except critical points of x_N. Also, since (12) coincides with (11) in U_1 and with (10) outside U, it follows that F' coincides with $\phi^{-1}(F_0)$ in U_1 and with F outside U, as required.

5-4. NEIGHBORHOOD OF A CRITICAL POINT

As already indicated, the first thing is to examine the special case of a neighborhood of the origin on a quadratic hypersurface (9). The information will then be transferred to M by the map ϕ^{-1} of the last section. Most of the argument for the main result of this section will be carried out in sequences of exercises.

Exercises. **5-5.** Let H be a hypersurface with equation $z = f(y_1, y_2, \ldots, y_n)$ in Euclidean $(n + 1)$-space. Prove that the orthogonal trajectories on H of the sections by the hyperplanes z = constant project onto the orthogonal trajectories in the n-space $z = 0$ of the family of sets f = constant.

(The point is to see that the tangent linear space to a horizontal section of H projects into the tangent linear space to a set f = constant and that a line orthogonal to the first projects into a line orthogonal to the second.)

5-6. Working now in the n-space with the coordinates $y_1, y_2, \ldots, y_n$, consider the family of hypersurfaces $f(y_1, y_2, \ldots, y_n)$ = constant. Show that the orthogonal trajectories to these satisfy the differential equations

$$\frac{dy_1}{f_1} = \frac{dy_2}{f_2} = \cdots = \frac{dy_n}{f_n} \tag{14}$$

where $f_i = \partial f/\partial y_i$.

Hence, show that for the family of hypersurfaces

$$y_1^2 + y_2^2 + \cdots + y_r^2 - y_{r+1}^2 - \cdots - y_n^2 = \text{constant}, \tag{15}$$

the equations of the orthogonal trajectories are

$$\begin{gathered} y_2'y_1 = y_1'y_2, \ \ldots, \ y_r'y_1 = y_1'y_r, \\ y_{r+2}'y_{r+1} = y_{r+1}'y_{r+2}, \ \ldots, \ y_n'y_{r+1} = y_{r+1}'y_n, \\ y_1y_{r+1}' = y_1'y_{r+1}, \end{gathered} \tag{16}$$

where $y_1', y_2', \ldots, y_n'$ are constants. In fact, the set of equations above is the orthogonal trajectory through the point $(y_1', y_2', \ldots, y_n')$.

Note that there is exactly one curve of the family (16) through each point $(y_1', y_2', \ldots, y_n')$ except $(0, 0, \ldots, 0)$. Note also that the member of the family (15) through the origin is an $(n-1)$-dimensional cone with vertex at the origin; thus, we would not expect the orthogonality condition to work at that point.

5-7. Note that all the Eqs. (16) except the last are linear, and in fact are linearly independent except when y_1' or y_{r+1}' is zero. Thus when $y_1' \neq 0$, $y_{r+1}' \neq 0$ the Eqs. (16) represent a hyperbola. Note that the condition on y_1 and y_{r+1}' is an accident of the numbering of the y_i. Show by suitable arrangement of the equations of the orthogonal trajectories that, so long as one of $y_1', y_2', \ldots, y_r'$ is nonzero and one of $y_{r+1}', y_{r+2}', \ldots, y_n'$ is nonzero, the trajectory through $(y_1', y_2', \ldots, y_n')$ is a hyperbola.

5-8. As a complement to the last exercise show that if one of the orthogonal trajectories has $y_i' = 0$ at any point, then $y_i = 0$ all along it. (Deduce this directly from the differential equations.) Use this to show that a trajectory containing a point with $y_1' = y_2' = \cdots = y_r' = 0$ is a straight line segment ending at the origin. Obtain a similar result for a trajectory containing a point with $y_{r+1}' = y_{r+2}' = \cdots = y_n' = 0$.

In addition, show that if we consider the orthogonal trajectories to the family (15) with $-1 \leq c \leq 1$, then the set of trajectories in the linear space $y_1 = y_2 = \cdots = y_r = 0$ forms an $(n-r)$-cell E^{n-r} whose boundary sphere S^{n-r-1} is in (15) with $c = -1$, while the trajectories satisfying $y_{r+1} = y_{r+2} = \cdots = y_n = 0$ form an r-cell E^r whose boundary sphere S^{r-1} is in (15) with $c = 1$. Note that the two cells E^r and E^{n-r} have only the origin in common.

Before going any further with the general discussion, it is useful to look more closely at the example with $n = 3$. Changing notation for convenience, consider the family of surfaces

$$x^2 - y^2 - z^2 = c$$

with $-1 \leq c \leq 1$. Note that the two surfaces obtained by putting c equal to -1 and 1 are hyperboloids of one sheet and two sheets, respectively, the two-sheeted surface being inside the one-sheeted one (cf. Fig. 5-2).

The orthogonal trajectories to this family of surfaces are all, as shown earlier, arcs of hyperbolas, except for a set of line segments forming the cells E^1 and E^2, as described in Exercise 5-8. The boundaries of these cells are marked in Fig. 5-2 as S^0 and S^1, respectively.

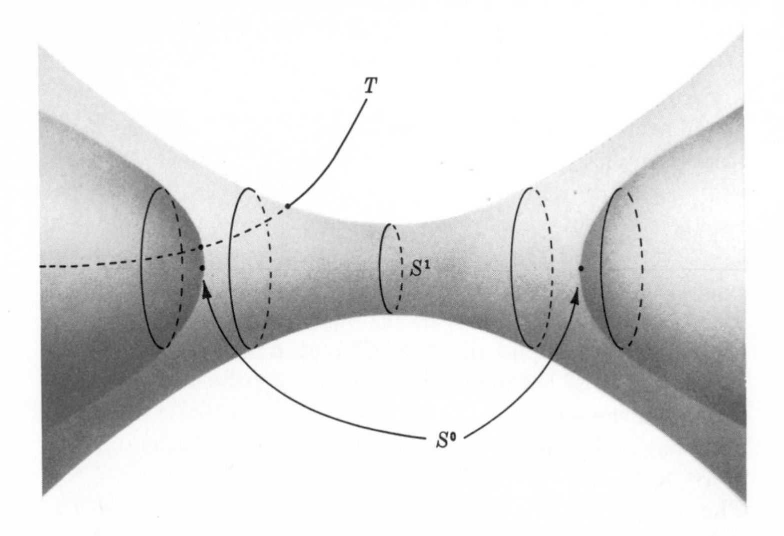

FIGURE 5-2 *An orthogonal trajectory T beginning in a neighborhood of S^0 ends in a neighborhood of S^1.*

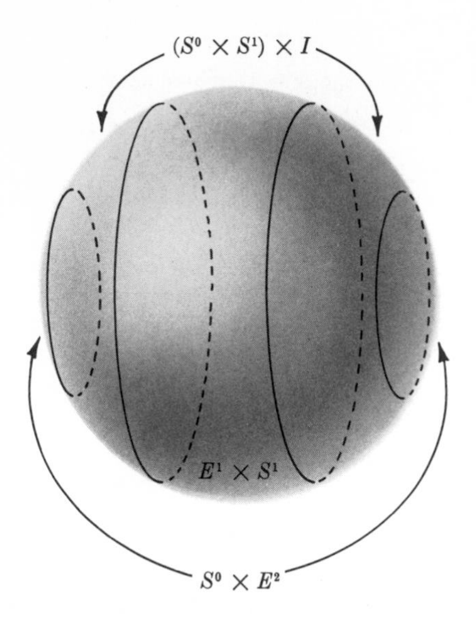

FIGURE 5-3 *S^2 decomposed into 3 sets obtained from Figure 5-2.*

Consider now those trajectories beginning in a neighborhood of S^0 and ending in a neighborhood of S^1. The neighborhood of S^0 in question is the union of two disks and that of S^1 is a circular strip around the one-sheeted hyperboloid. These neighborhoods can be represented as $S^0 \times E^2$ and $E^1 \times S^1$, respectively. Also, it can be seen intuitively that if these two neighborhoods are taken to be the right size, then the orthogonal trajectories beginning in one are exactly those ending in the other. And the union of all these arcs, along with the cells E^1 and E^2, forms a 3-cell. The boundary of this cell is a 2-sphere decomposed as indicated in Fig. 5-3. That is, it consists of the union of $S^0 \times E^2$ and $E^1 \times S^1$ (the above-mentioned neighborhoods of S^0 and S^1 on the two hyperboloids) and the set shaded in the figure. The latter is the union of orthogonal trajectories joining the points of the boundaries of $S^0 \times E^2$ and $E^1 \times S^1$ (both boundaries being $S^0 \times S^1$). This set can thus be represented as $(S^0 \times S^1) \times I$, where I is a line segment.

The following sequence of exercises aims at generalizing the description just given.

Exercises. **5-9.** Let S^{n-1} be the sphere

$$x_1^2 + x_2^2 + \cdots + x_n^2 = 1$$

in n-space. Let S^{r-1} be the $(r-1)$-sphere on S^{n-1} given by

$$x_1^2 + x_2^2 + \cdots + x_r^2 = 1,$$
$$x_{r+1} = x_{r+2} = \cdots = x_n = 0,$$

and let S^{n-r-1} be the $(n-r-1)$-sphere on S^{n-1} given by

$$x_1 = x_2 = \cdots = x_r = 0,$$
$$x_{r+1}^2 + x_{r+2}^2 + \cdots + x_n^2 = 1.$$

Check that any point p on S^{r-1} and any point q on S^{n-r-1} are joined by a unique great circle arc that is a quarter circumference. That is, the directions of p and q from the center of S make an angle of $\pi/2$. Also show that two such arcs pq and $p'q'$ have no point in common, except possibly an end point, if $p = p'$ or $q = q'$.

5-10. Fix p on S^{r-1}, and for any great circle arc pq with q on S^{n-r-1} (as in the last exercise) let q_1 be the midpoint. Show that, as q varies throughout S^{n-r-1}, p being fixed, the union of the arcs pq_1 is an $(n-r)$-cell. Hence, show that the set of points on S^{n-1} at angular distance $\leq \pi/4$ from S^{r-1} is of the form $S^{r-1} \times E^{n-r}$ where E^{n-r} is an $(n-r)$-cell.

5-11. Deduce from the last exercise that S^{n-1} is the union of $S^{r-1} \times E^{n-r}$ and $E^r \times S^{n-r-1}$. Note that these sets have a common boundary in S^{n-1}, namely, $S^{r-1} \times S^{n-r-1}$. Another way of saying all this is, S^{n-1} is the union of the spaces $S^{r-1} \times E^{n-r}$ and $E^r \times S^{n-r-1}$, the point (p, q) in $S^{r-1} \times S^{n-r-1} = S^{r-1} \times$ (boundary of E^{n-r}) being identified with (p, q) in $S^{r-1} \times S^{n-r-1} =$ (boundary of E^r) $\times S^{n-r-1}$.

5-12. There is a variation of the last exercise. Namely, show that S^{n-1} is the union of three sets A, B, C where A, the set of points at angular distance $\leq \pi/6$ from S^{r-1}, is of the form $S^{r-1} \times E^{n-r}$; B, the set of points at angular distance $\leq \pi/6$ from S^{n-r-1}, is of the form $E^r \times S^{n-r-1}$; while C, the set at angular distance $\geq \pi/6$ from S^{r-1} and $\leq \pi/6$ from S^{n-r-1}, is of the form $(S^{r-1} \times S^{n-r-1}) \times I$, where I is a line segment, say the unit interval $(0, 1)$. Here the point (p, q) on A with q in $S^{n-r-1} =$ boundary of E^{n-r} is identified with $(p, q, 1)$ in C and (p, q) in B with p in $S^{r-1} =$ boundary of E^r, is identified with $(p, q, 0)$ in C.

Note that this decomposition of S^{n-1} corresponds to the decomposition of S^2 described earlier (Fig. 5-3), except that there S^2 appeared not as the surface of the unit sphere, but as a set associated with a family of quadratic surfaces. This situation will now be reproduced in the general case.

5-13. Consider the family of quadratic hypersurfaces in n-space

$$x_1^2 + x_2^2 + \cdots + x_r^2 - x_{r+1}^2 - \cdots - x_n^2 = c \tag{17}$$

with $-1 \leq c \leq 1$.

Note that the spheres S^{r-1} and S^{n-r-1} of Exercise 5-9 lie on the members of this family with $c = 1$ and $c = -1$, respectively. Also check that the direction from the origin to any point of S^{r-1} or S^{n-r-1} makes an angle $\geq \pi/4$ with any line in the cone

$$x_1^2 + x_2^2 + \cdots + x_r^2 - x_{r+1}^2 - \cdots - x_n^2 = 0.$$

This implies that the sets A and B of Exercise 5-12 project homeomorphically from the origin into the hypersurfaces (17) with $c = 1$ and $c = -1$, respectively. Denote this projection by π. Let p and q be points of the boundaries of A and B, respectively, that are joined by one of the great circle arcs, say α, of which C is composed. Verify that $\pi(p)$ and $\pi(q)$ are on a hyperbolic arc β, one of the orthogonal trajectories of the family (17), and that α projects onto β from the origin.

Thus there is a homeomorphic image of S^{n-1} under π consisting of the union of $\pi(A)$ and $\pi(B)$ on (17) with $c = 1$ and $c = -1$, respectively, and $\pi(C)$, a union of arcs of orthogonal trajectories of the family (17).

5-14. Verify that the last two exercises imply the existence of a differentiable function f on the n-cell E^n (whose boundary is S^{n-1}) such that $f = 0$ on A, $f = 1$ on B, and $f = t$ on the subset $S^{r-1} \times S^{n-r-1} \times \{t\}$ of $C = S^{r-1} \times S^{n-r-1} \times I$.

5-5. NEIGHBORHOOD OF A CRITICAL LEVEL; SUMMING UP

All the results necessary for describing the neighborhood of a critical level of a differentiable function have now been obtained, and will simply be put together in this section.

Theorem 5-1. *Let M be a differentiable manifold of dimension n and let f be a differentiable function and, in the notation of Definition* 5-1, *let M_c be a critical level of f with one nondegenerate critical point P on it. Let M_a and M_b be noncritical levels of f with $a < c < b$, M_c being the only critical level between them. Then P has a neighborhood E^n on M, an n-cell whose boundary S^{n-1} is the union of three sets A, B, C. A is on M_a and is diffeomorphic to $S^{r-1} \times E^{n-r}$ (for some r), B is on M_b and is diffeomorphic to $E^r \times S^{n-r-1}$, while C, lying between M_a and M_b, can be expressed as $S^{r-1} \times S^{n-r-1} \times I$. Here a point (p, q) of A with $p \in S^{r-1}$ and $q \in S^{n-r-1}$ = boundary of E^{n-r} is identified with $(p, q, 0)$ on C, and a point (p, q) of B with p in S^{r-1} = boundary of E^r and q in S^{n-r-1} is identified with $(p, q, 1)$ in C.*

In addition, if the cell E^n is removed from the part of M between M_a and M_b, the remainder can be represented as $(M_a - A) \times I$ where $(M_a - A) \times \{0\}$ is identified with $M_a - A$ and $(M_a - A) \times \{1\}$ with $M_b - B$.

Proof. It follows from Exercise 5-4 that it is only necessary to prove the present theorem for values of a and b near c; just how near will be determined in the course of the proof.

Start by taking a neighborhood U of P, as in Section 5-3, and construct the modified family F' of orthogonal trajectories of the level sets of f. Remember that in a smaller neighborhood U' of P, these coincide with $\phi^{-1}(F_0)$ where ϕ is as in Section 5-3 and F_0 denotes the family of orthogonal trajectories of the horizontal sections of a hypersurface

$$z = \Sigma c_i y_i^2$$

in $(n + 1)$-space with the c all ± 1. By Exercise 5-5, projecting on the space $z = 0$, F_0 projects on the orthogonal trajectories of the family

$$\Sigma c_i y_i^2 = c. \tag{18}$$

If necessary, by a change of scale in the y_i it can be assumed that the cell E^n described in Exercise 5-14 is in the image of U' under ϕ. The inverse image of this cell in M then gives the cell E^n described in the statement of this theorem, provided that M_a and M_b are taken as the noncritical levels that, in U', are mapped by ϕ into (18) with c equal to -1 and 1, respectively. The properties described in Exercise 5-13 then carry over to M showing that the boundary of E^n has all the properties stated in the present theorem.

The arcs of the family F' beginning on $M_a - A$ (and so, by the foregoing, ending on $M_b - B$) can then be used, as in Exercise 5-2, to check that the part of M between M_a and M_b after the removal of the cell E^n is $(M_a - A) \times I$. This completes the proof of the theorem.

Some further information on the present situation can be extracted from Exercise 5-9: that exercise implies that there is a sphere S^{r-1} on M_a and a sphere S^{n-r-1} on M_b such that members of F' starting on $M_a - S^{r-1}$ end on $M_b - S^{n-r-1}$ (and the other way round), while the members of F' through points of S^{r-1} or S^{n-r-1} all end at P. Thus M contains two cells E^r and E^{n-r} having boundaries on M_a and M_b, respectively, and having just the point P in common.

Note, in addition, that S^{r-1} has a neighborhood in M_a of the form $S^{r-1} \times E^{n-r}$ and S^{n-r-1} in M_b has a neighborhood of the form $S^{n-r-1} \times E^r$.

6

Spherical Modifications

6-1. INTRODUCTION

The object of this chapter is to look at the material of the preceding one from a different point of view. In the last chapter the starting point was a compact differentiable manifold M with boundary $M_0 \cup M_1$. A differentiable function f was given on M, or alternatively an embedding of M in a Euclidean space was given, and neighborhoods of critical and non-critical levels were studied. Here attention is to be paid rather to the way in which the level manifolds of f vary, starting from M_0 and ending with M_1.

It was seen in Section 5-1 that, as c increases from 0 to 1, the level manifold M_c remains the same, topologically speaking, except when c crosses a critical level. Thus to get from M_0 to M_1 through the family of levels of f, we perform a finite number of operations, one corresponding to each critical level. Each operation transforms M_a just below a critical level (thinking here of the levels as horizontal sections of M suitably embedded in Euclidean space) into M_b just above that level. The operation in each case consists in removing a neighborhood

(A in the notation of Theorem 5-1) of a sphere embedded in M_a and replacing it by a neighborhood (namely, B) of a sphere of different dimension. The main object of this chapter, then, is to study this kind of operation on manifolds.

6-2. DIRECT EMBEDDING

As just indicated, the definitions about to be given are based exactly on the result of Theorem 5-1, but the notation is changed. The manifold M discussed here corresponds to a noncritical level of the function f on the manifold of that theorem.

So let M be a differentiable manifold of dimension n and let N be a submanifold of dimension r. Take a point p on N and a local coordinate system around p satisfying the conditions of Definition 3-1. Suppose that the corresponding coordinate neighborhood U (or rather its image in Euclidean space) is specified by inequalities $|x_i| \leq \delta(i = 1, 2, \ldots, n)$; U can be thought of as a topological product $V \times F$ where V is $U \cap N$, namely, a neighborhood of p in N, and F is a neighborhood of the origin in a Euclidean $(n - r)$-space, that is, a space in which $x_{r+1}, x_{r+2}, \ldots, x_n$ are the coordinates, F being specified by inequalities $|x_i| \leq \delta$, $i = r + 1, r + 2, \ldots, n$. The sets in $V \times F$ of the form $\{q\} \times F$, with q in V, are $(n - r)$-cells cutting across N and each meeting N at just one point. Thus U can be thought of as a union of slices, each slice being an $(n - r)$-cell cutting N at one point. A union of coordinate neighborhoods of the type just described will be called a tubular neighborhood of N in M. It can be shown (but this will not be needed here) that the entire tubular neighborhood can be expressed as a union of $(n - r)$-cells each cutting across N at one point; that is, the slices of overlapping coordinate neighborhoods can be made to coincide in the overlap.

Here one case is of special interest, namely, that in which a tubular neighborhood of N is diffeomorphic to $N \times F$ where F is an $(n - r)$-cell.

Definition 6-1. When this happens, N will be said to be *directly embedded in M.*

Examples

6-1. Let N be the circle $x^2 + y^2 = 1$, $z = 0$ in Euclidean 3-space, and let M be the 3-space itself. N is certainly a submanifold of M. Now let B be a solid torus with N as its center line (see Fig. 6-1). Let F be the section of B by the plane $y = 0$ and lying in the half space $x > 0$. F is a circular disk, that is, a 2-cell. Now take any point p in B, and take a plane through p and the z axis. This plane makes an angle θ with the (x, z) plane. θ can, of course, be taken as coordinate on N. Also draw a circle C through p parallel to the (x, y) plane, meeting F in q. Thus to p in B there corresponds a pair (θ, q). It is not hard to see that this expresses B as the product $N \times F$. Hence, N is directly embedded in M.

6-2. An example will now be given of a submanifold that is not directly embedded. Let M be the projective plane (Example 2-6). It was seen that this space can be represented as a hemisphere with the pairs of diametrically opposite points on the edge identified. As N, take a circle obtained by join-

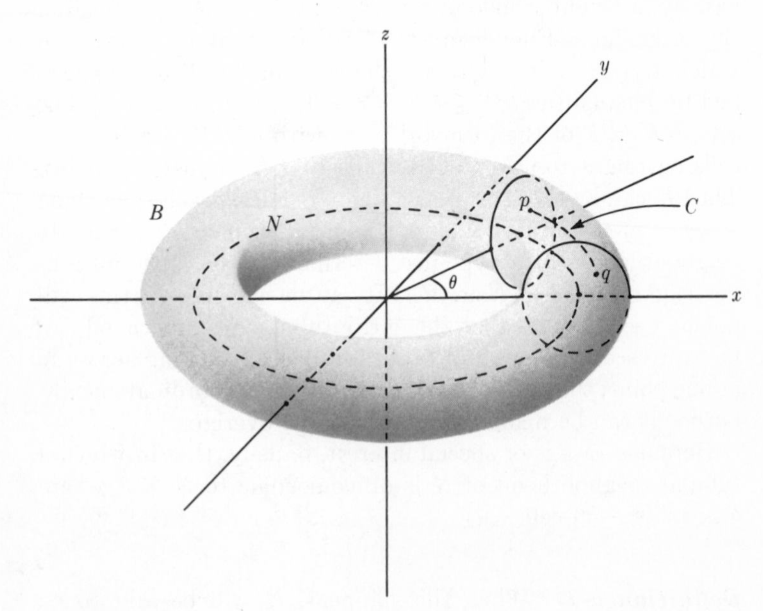

FIGURE 6-1 *Tubular neighborhood of S^1 and E^3 as a product.*

ing the ends of a great semicircle on the hemisphere. It is easy to see that N is a submanifold of M. A tubular neighborhood B of N in M can be constructed first on the hemisphere, where it appears as a strip having N as center line. Certainly B can be expressed as a union of slices, 1-cells cutting across N, that is, arcs of circles on the hemisphere at right angles to N. However, the identification of diametrically opposite points means that, to construct B, the strip on the hemisphere is given a half twist before the ends are joined up. Thus B is a Möbius strip. It can be shown that this is not homeomorphic to a product $N \times F$, where F is a 1-cell. So N is not directly embedded in M.

The last example illustrates a point that is worth pursuing further, as it leads to an important distinction between two kinds of manifolds. The existence (as in Example 6-2) of a nondirectly embedded circle means that, as we travel along such a circle, returning to the starting point, the manifold develops a kind of twist. Thus a neighborhood of the circle in Example 6-2 is a twisted strip. Such a twisting cannot happen if every circle in the manifold is directly embedded.

Definition 6-2. Let M be a differentiable manifold. If every circle embedded in M as a submanifold is directly embedded, M will be called *orientable*, otherwise *nonorientable*. This definition, equivalent to others more usually given in the literature, is appropriate here because this is the way the concept is to be used.

Example

6-3. The sphere is orientable but the projective plane is not.

It can be shown (cf. Chapter 7) that the compact two-dimensional manifolds can be completely classified. That is, each such manifold is homeomorphic to a sphere with p handles (for some p) or to a sphere with k holes (for some k), the diametrically opposite points on the circumference of each hole being identified. The torus is an example of the first kind, with $p = 1$, and the projective plane is an example of the second kind, with $k = 1$. The manifolds of the first kind are all orientable and those of the second kind are nonorientable. The second statement is easy to see, since an arc on the k-holed sphere joining diametrically opposite points on one of the holes becomes, after the identification, a nondirectly embedded circle. To see that a sphere with p handles is orientable, let S be a circle on it, embedded as a submanifold, and let B be a tubular neighborhood of S. Fix a direction along S, and at p on S take the tangent line pointing in that direction. Then take a tangent to the surface at p, with a direction marked on it, perpendicular to S at p and making a right-handed system with the directed tangent to S and the outward normal of the surface. This defines at each p of S a positive direction in B at right angles to S, and so defines B as a product $S \times I$.

It can be shown (but this will not be done here) that the definition of orientability can be formulated in terms of local coordinate systems. That is, a manifold M is orientable if and only if it has a covering by local coordinate neighborhoods such that, for any two that overlap, the Jacobian determinant of the corresponding coordinate transformations is positive.

***Exercise.* 6-1.** Show that the six coordinate neighborhoods on S^2 in Example 2-6 satisfy the condition just stated, provided that the pair of coordinates in each neighborhood is taken in the right order.

6-3. DEFINITION OF MODIFICATIONS

The notion of spherical modification can now be described. Let M be an n-dimensional differentiable manifold and suppose that S^r is an r-dimensional sphere that is a directly embedded submanifold of M. Briefly, S^r will be called a directly embedded r-sphere. Thus S^r has a neighborhood in M that is diffeomorphic to $S^r \times E^{n-r}$ where E^{n-r} is an $(n-r)$-cell. Now the boundary of B is the manifold $S^r \times S^{n-r-1}$; thus, if the interior of B is removed from M, a manifold with boundary is obtained, the boundary being $S^r \times S^{n-r-1}$. On the other hand, $S^r \times S^{n-r-1}$ is also the boundary of the differentiable manifold $E^{r+1} \times S^{n-r-1}$. And so, as explained in Section 2-7, the union of $M - \text{Int } B$ and $E^{r+1} \times S^{n-r-1}$ can be formed with identification of the boundaries. In fact, this can be done so that the result is a differentiable manifold M'.

Definition 6-3. M' will be said to be obtained from M by a *spherical modification of type* r (the dimension of the sphere whose neighborhood was removed).

Examples

6-4. Let M be a 2-sphere and take a zero-dimensional sphere S^0 in M (see Fig. 6-2). S^0 has a neighborhood consist-

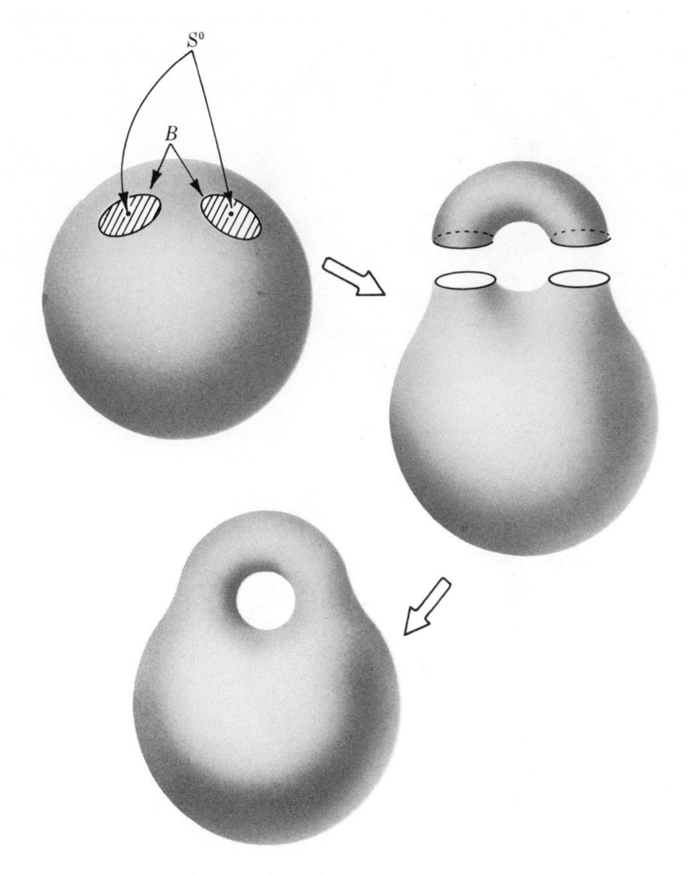

FIGURE 6-2 *Torus by modification out of sphere.*

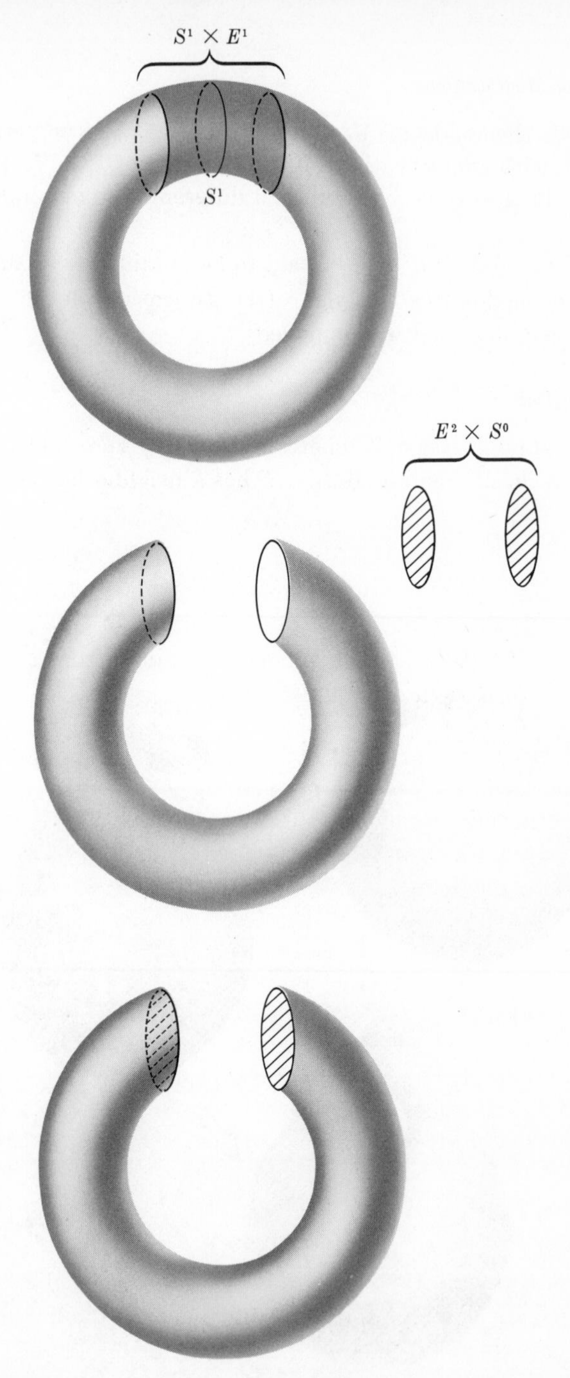

FIGURE 6-3 *Sphere by modification out of torus; the sphere appears in the third diagram as a cylinder with the ends closed by disks.*

ing of two disjoint disks. This is certainly of the form $S^0 \times E^2$, and so S^0 is directly embedded. This neighborhood is to be taken as B. $M - \text{Int } B$ is a sphere with two holes in it. $E^1 \times S^1$ is a cylinder, and when its ends are attached to the circumferences of the two holes, the resulting surface is a sphere with one handle, in other words, a torus. Thus the torus is obtained from the 2-sphere by a spherical modification of type 0.

6-5. As another example of a spherical modification, consider the reverse operation: that is, take M to be the torus, S^1 a circle as shown in Fig. 6-3. S^1 has a neighborhood $S^1 \times E^1$ in M, namely, a strip wrapped around the torus. When the interior of this strip is removed, the remainder has boundary $S^1 \times S^0$. This is also the boundary of $E^2 \times S^0$, a pair of disjoint disks, and if these are inserted, the result is a 2-sphere M'. Thus the 2-sphere is obtained from the torus by a spherical modification of type 1.

The situation illustrated by the foregoing examples, transforming a sphere into a torus and then back into a sphere, is quite general. That is, if M' is obtained from M by a spherical modification, then M can be obtained from M' by a spherical modification. For suppose, as in Definition 6-3, that M' is obtained from M by removing the tubular neighborhood $S^r \times E^{n-r}$ of S^r, replacing it by $E^{r+1} \times S^{n-r-1}$. Now $E^{r+1} \times S^{n-r-1}$ contains the sphere $\{p_0\} \times S^{n-r-1}$, where p_0 is some interior point of E^{r+1}. It follows that M' contains the sphere $\{p_0\} \times S^{n-r-1}$ and that this sphere has the tubular neighborhood $E^{r+1} \times S^{n-r-1}$ in M', so that it is directly embedded. And so M is obtained from M' by removing this tubular neighborhood of $\{p_0\} \times S^{n-r-1}$, replacing it by $S^r \times E^{n-r}$. This is a spherical modification of type $n - r - 1$.

Example

6-6. A general type of example is obtained from Theorem 5-1: If W is a differentiable manifold and f a differentiable function on it and if M and M' are level manifolds of f separated by one critical level, then M' is obtained from M by a spherical modification.

6-4. THE TRACE OF A MODIFICATION

Example 6-6 shows how a function with a critical point gives rise to a spherical modification. It will now be shown that every spherical modification can be obtained in exactly this way. The idea is to take a pair of manifolds related by a spherical modification and to assemble a manifold with boundary and construct on it a function that will have the given manifolds as noncritical levels separated by a critical level. The pattern for this construction is given by the knowledge of what the neighborhood of a critical level should look like.

As a guide to understanding the general construction, consider the following example.

Example

6-7. In Example 6-4 it was seen that a torus M_1 can be obtained from a sphere M_0 by a modification of type 0. The set Int B_0 that is removed from M_0 consists of two disjoint open disks, so that $M_0 - \text{Int } B_0$ is a sphere with two holes. For the present purpose it is more convenient to think of $M_0 - \text{Int } B_0$ as the surface of a cylinder, bent round as shown in Fig. 6-4; $(M_0 - \text{Int } B_0) \times I$ is then a thickened cylinder. As Fig. 6-4 indicates, the ends of this cylinder are unions of radial line segments, each segment being a set of the form $\{p\} \times I$, where p is in the boundary of $M_0 - \text{Int } B_0$. The union of the ends of the cylinder can, of course, be expressed as $(S^0 \times S^1) \times I$. The inner surface of the cylinder, namely, $(M_0 - \text{Int } B_0) \times \{0\}$ is to be thought of as $M_0 - \text{Int } B_0$, and the outer surface, $(M_0 - \text{Int } B_0) \times \{1\}$, as $M_1 - \text{Int } B_1$. B_0 is a pair of disks, whereas B_1 (cf. Example 6-4) is a cylinder $E^1 \times S^1$. So the inner surface of $(M_0 - \text{Int } B_0) \times I$ can be made into a sphere by replacing B_0, while the outer surface can be made into a torus by adding B_1.

On the other hand, consider a 3-cell E^3 with its boundary S^2 decomposed into the three sets $S^0 \times E^2$, $S^1 \times E^1$, and $(S^0 \times S^1) \times I$, in fact, the A, B, C of Exercise 5-12 (see Fig. 6-5). Note that these sets can be identified, respectively, with

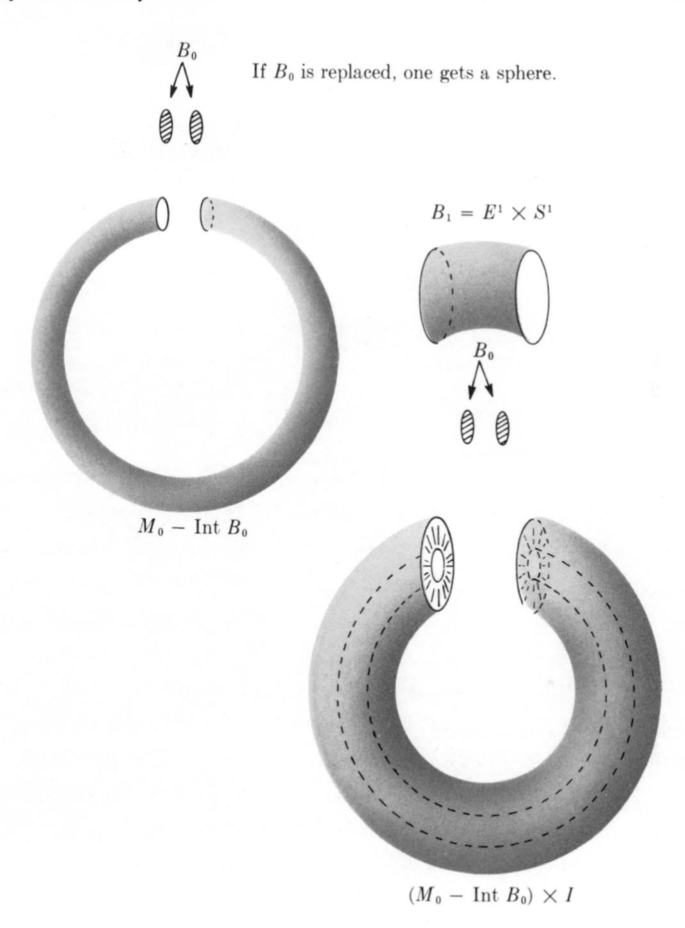

FIGURE 6-4 *B_0 added to the inner surface of $(M_0 - Int\, B_0) \times I$ gives a sphere, B_1 added to the outer surface gives a torus.*

B_0, B_1, and the ends of the thick cylinder $(M_0 - \text{Int } B_0) \times I$. Moreover, the radial line segments on the ends of the cylinder can be identified in a one-to-one manner with the great circle arcs forming $(S^0 \times S^1) \times I$ on the surface S^2. So if the set $(S^0 \times S^1) \times I$ on the surface of the cell E^3 is identified with the corresponding set on the ends of the thick cylinder, a solid M is obtained, with B_0 and B_1 automatically falling into place to make an inner boundary M_0 and an outer boundary M_1.

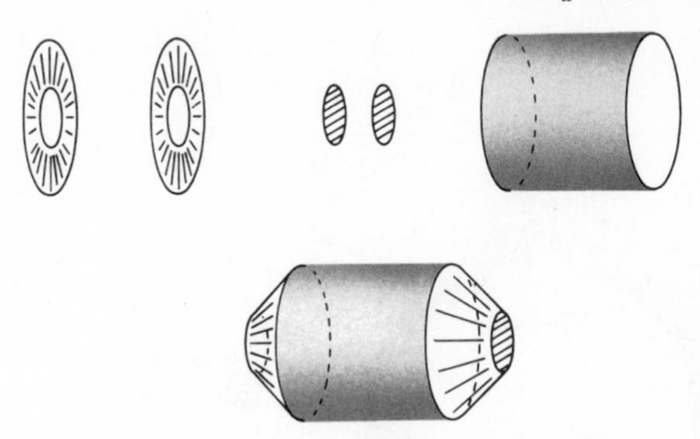

FIGURE 6-5 *The sets B_0, B_1, and $(S^0 \times S^1) \times I$ put together to form a 2-sphere, the boundary of a 3-cell.*

Another piece of information comes out of the preceding construction: it was seen in Exercise 5-14 that there is a differentiable function f on the 3-cell E^3 with the following properties. On the boundary S^2 of E^3, $f = 0$ on the set $B_0 = S^0 \times E^2$, $f = 1$ on the set $B_1 = E^1 \times S^1$, and on the subset $(S^0 \times S^1) \times \{t\}$ of $(S^0 \times S^1) \times I$, $f = t$. Otherwise f takes values between 0 and 1 on E^3 and has just one nondegenerate critical point at the center. Clearly, f can be extended to the whole of M, as constructed in the last paragraph, by setting $f = t$ on the points of $(M_0 - \text{Int } B_0) \times \{t\}$ in $(M_0 - \text{Int } B_0) \times I$. If it has been arranged already that $(M_0 - \text{Int } B_0) \times I$ and E^3 have been put together to form a differentiable manifold, then f will be a differentiable function equal to 0 on M_0 and to 1 on M_1, and with just one nondegenerate critical point at the center of E^3.

With this example in mind, the corresponding construction can now be described in the general case. Suppose that M_0 is a differentiable manifold of dimension n and that M_1 is obtained from it by a spherical modification of type r. The modification is carried out by removing from M_0 a set B_0 diffeomorphic to $S^r \times E^{n-r}$, a tubular neighborhood of the directly embedded sphere S^r, and then replacing it by $B_1 = E^{r+1} \times S^{n-r-1}$. To construct M, following the pattern of the preceding example, start with $(M_0 - \text{Int } B_0) \times I$. Part of the boundary of this is of the form $(S^r \times S^{n-r-1}) \times I$. On the other hand, the boundary S^n of an n-cell E^{n+1} can be expressed as the union $A \cup B \cup C$ (cf. Exercise 5-12), where $A = S^r \times E^{n-r}$, $B = E^{r+1} \times S^{n-r-1}$, and $C = (S^r \times S^{n-r-1}) \times I$. Form the union now of $(M_0 - \text{Int } B_0) \times I$ and E^{n+1}, identifying the subsets $S^r \times S^{n-r-1} \times I$ appearing in both. This automatically replaces $A = B_0$ in $M_0 - \text{Int } B_0$ to form M_0 and $B = B_1$ in $M_1 - \text{Int } B_1$ to form M_1, and so forms a manifold M having the disjoint union of M_0 and M_1 as boundary.

Moreover, a function f is to be constructed on M as in the three-dimensional example. Take a function f on E^{n+1}, as in Exercise 5-14, such that on the boundary sphere S^n, $f = 0$ on A, $f = 1$ on B, and on C, f has the value t on the set $S^r \times S^{n-r-1} \times \{t\}$. Also, f has exactly one nondegenerate critical point at the center of E^{n+1}. Then f can be extended to all of f by setting it equal to t on the set $(M_0 - \text{Int } B_0) \times \{t\}$ in $(M_0 - \text{Int } B_0) \times I$.

If $(M_0 - \text{Int } B_0) \times I$ and E^{n+1} have been put together properly, M will be a differentiable manifold and f will be a differentiable function. On $(M_0 - \text{Int } B_0) \times I$ the value of f, namely t, the parameter on I, can always be taken as one of the local coordinates, and so, by Lemma 4-3, f has no critical point there. In other words, its only critical point is the nondegenerate one with type number $r + 1$ at the center of E^{n+1}.

The result obtained can be summed up as follows.

Theorem 6-1. *Let M_1 be obtained from M_0 by a spherical modification of type r. Then there is a differentiable manifold M whose boundary is the disjoint union of M_0 and M_1, and a differentiable function f on M, equal to 0 on M_0, equal to 1 on M_1, otherwise having values between 0 and 1 and having exactly one nondegenerate critical point with type number $r + 1$.*

A finite number of applications of this theorem gives the following result.

Theorem 6-2. *Let M_1 be obtained from M_0 by a finite number of spherical modifications. Then there is a differentiable manifold M whose boundary is the disjoint union $M_0 \cup M_1$ and a* *differentiable function f on M with value 0 on M_0, 1 on M_1, otherwise with values between 0 and 1 and with a nondegenerate critical point corresponding to each modification.*

Definition 6-4. The manifold M constructed in this theorem will be called the *trace of the sequence of modifications* transforming M_0 into M_1.

Another piece of terminology is useful. If a modification ϕ leading from M_0 to M_1 consists in removing a neighborhood of a sphere S^r in M_0; then, looking at the trace of ϕ and at the orthogonal trajectories to the level sets of the corresponding function f (in the notation of Theorem 6-1), it will be seen (cf. Exercise 5-8) that the trajectories starting at points of S^r all end at the critical point of f. Thus as we go through the levels of f from M_0 to M_1, S^r shrinks to a point along the orthogonal trajectories. And then, as we continue beyond the critical level, the sphere S^{n-r-1} appears, expanding from the critical point along the orthogonal trajectories till it reaches M_1. And so it is convenient to speak of ϕ as *shrinking* S^r *and introducing* S^{n-r-1}.

6-5. COBOUNDING MANIFOLDS

There is another way of expressing Theorem 6-2, in terms of a relation between manifolds known as cobounding.

Definition 6-5. Two compact differentiable manifolds M_0 and M_1 are said to *cobound* or to be *cobounding* if there is a compact differentiable manifold M such that the boundary of M is the disjoint union $M_0 \cup M_1$. In particular, if M_1 is empty, M_0 is called a bounding manifold.

Examples

6-8. A sphere S^n is a bounding manifold, since it is the boundary of E^{n+1}.

6-9. If M_0 is a sphere with p handles, it is a bounding manifold, for it is the boundary of the solid sphere with p solid handles.

These examples may seem rather trivial, but it is hard to give significant examples. In general, the tests that must be applied to a pair of manifolds to see if they are cobounding are very complicated, and will not be described here.

On the other hand, the following result can be stated.

Theorem 6-3. *If M_0 and M_1 are compact differentiable manifolds, then they are cobounding if and only if each can be obtained from the other by a finite number of spherical modifications.*

Proof. If M_0 and M_1 are given cobounding, that is, if it is given that their disjoint union is the boundary of a manifold M, then by Theorem 6-2 there is a function f on M equal to 0 and 1 on M_0 and M_1, respectively, with just a finite number of critical points, all nondegenerate and all on different critical levels. Consider the level sets M_c of f as c increases from 0 to 1. By Exercise 5-4, M_c remains topologically the same until a critical level is passed, and then (Theorem 5-1) it is changed by a spherical modification. This happens just a finite number of times on the way from M_0 to M_1.

Conversely, if M_1 is obtained from M_0 by a finite number of spherical modifications, Theorem 6-2 says that M_0 and M_1 are cobounding.

6-6. DISPLACEMENT AND ISOTOPY

If the definition of a spherical modification is examined, it appears that the operation depends on the sphere S^r to be shrunk and also on the expression, as a product, of a tubular neighborhood of S^r in the given manifold M_0. On the other hand, the first step in the construction of the modification is the removal of the tubular neighborhood of S^r. So another sphere S_1^r with the same tubular neighborhood as S^r would give rise to the same modification. This will happen, for example, if S_1^r is obtained from S^r by a small displacement such that S_1^r meets each cellular slice of the product $S^r \times E^{n-r}$ (the tubular neighborhood of S^r) in one point. $S^r \times E^{n-r}$ can then automatically be expressed as $S_1^r \times E^{n-r}$, and so the same modification is obtained whether we start with S^r or S_1^r. That is, the result M_1 of applying the modification shrinking S_1^r is the same as that of the modification shrinking S^r. Also, it is not hard to see that the traces of these modifications are the same.

Here it will be noticed that the product structure on the tubular neighborhood of S_1^r is determined by the corresponding structure for S^r. It is important to note that, even without changing the sphere S^r, a change in the product structure of the tubular neighborhood of S^r may change the result of the corresponding modification. This is illustrated by the following simple example.

Example

6-10. Take the sphere S^2 and the 0-sphere S^0 on it as in Example 6-4. Now there are actually two ways of expressing a neighborhood of S^0 as a product $S^0 \times E^2$, one obtained from the other by reversing the sense of rotation on one of the disks. This means that, when $S^1 \times E^1$ is attached with the appropriate identifications on the boundaries, the identification can be done in two ways. One (the orientable way) is as described in Example 6-4, the result being a torus. The other (the nonorientable way) will give the one-sided Klein surface (cf. Fig. 7-22).

Exercises. **6-2.** Let S^r be a directly embedded sphere in a manifold M and let B be a tubular neighborhood of S^r. If two expressions of B as a product $S^r \times E^{n-r}$ are given, this corresponds to giving a diffeomorphism f of B on itself. That is, the point (p, q) in one product representation of B is mapped by f on the point (p, q) in the other representation. Prove that if f can be extended to a diffeomorphism of M on itself, the modifications corresponding to the two expressions will have the same result and the same trace.

6-3. As a variation of the last exercise, suppose that S^r is directly embedded in M and that B_1 and B_2 are two tubular neighborhoods of S^r, both expressed as $S^r \times E^{n-r}$. Thus, as in the foregoing, there is a map f of B_1 on B_2. Prove that if f can be extended to a diffeomorphism of M on itself, the modifications constructed using B_1 and B_2 have the same result and the same trace.

6-4. There is a special case of the last exercise that is important. Let S^r have tubular neighborhoods $B_1 \subset B_2 \subset B_3$, where $B_i = S^r \times E_i$, $i = 1, 2, 3$, and the E_i are solid spheres with center the origin in some

$(n - r)$-space, and $E_1 \subset E_2 \subset E_3$. Construct a diffeomorphism of $(n - r)$-space on itself that is the identity outside E_3 and that maps E_2 onto E_1.

(*Hint:* Construct the map by shifting points radially inward to the origin, the distance of shift being determined by a function like that of Example 2-4.)

Hence, show that there is a diffeomorphism of M onto itself mapping B_2 onto B_1.

Combined with Exercise 6-3, the point of Exercise 6-4 is that, in performing a modification shrinking a sphere S^r, the tubular neighborhood of S^r that is used can be taken arbitrarily small.

There are some more general forms of the results just described, but the details will not be given here. The most important situation is that of two directly embedded spheres S_1 and S_2 that are the images of isotopic maps of S^r into M. Here in effect we can think of S_2 as being obtained from S_1 by a large displacement that is made up of a sequence of small displacements, as described at the beginning of the section. It can be shown that a given product structure on a tubular neighborhood of S_1 will induce a product structure on a tubular neighborhood of S_2 such that the corresponding modifications shrinking S_1 and S_2 give the same results and have the same trace. This can be seen, for example, by repeated application of the corresponding result on small displacements.

There is one other point in this connection that must be examined; that is, a comparison must be made between the traces of two sequences of modifications where the spheres shrunk in one sequence are obtained from those shrunk in the other by small displacements. The thing to notice is that it is not sufficient to look at the traces of the individual modifications; attention must also be paid to the way in which they are put together to form the traces of the sequences.

For the present purpose it is sufficient to look at the following special case. Let ϕ be a modification transforming a manifold M_0 into M_1, the construction starting with the removal of a tubular neighborhood B of a directly embedded sphere S in M_0. Let ϕ' be a second modification on M_0, which starts by removing a tubular neighborhood B' of a sphere S', and suppose that there is a continuous map $F: M_0 \times I \to M_0$

such that F restricted to $M_0 \times \{0\}$ acts as the identity of M_0 on itself; restricted to $M_0 \times \{1\}$ it carries $B' \times \{1\}$ onto B, with the appropriate product structure; and restricted to each set $M_0 \times \{t\}$ it is a homeomorphism. Write g for the restriction of F to $M_0 \times \{1\}$. Then, using the method of Exercise 6-3 there is a homeomorphism G of the trace W of ϕ onto the trace W' of ϕ' whose restriction to M_0 is g^{-1}. The idea now is to adjust G so that its restriction to M_0 is the identity.

To do this define a map H of W' on itself as follows. Noting that, since M_0 is a noncritical level of a function on W' (cf. Theorem 6-2) it has a neighborhood of the form $M_0 \times I$ in W', with $M_0 \times \{0\}$ identified with M_0 (Exercise 5-2). Define H to be the identity outside this neighborhood, and on this neighborhood $M_0 \times I$ define

$$H(p, t) = F(p, 1 - t).$$

Note that $H(p, 1) = F(p, 0) = p$ and so the definitions of H on $M_0 \times I$ and outside $M_0 \times I$ fit together to form a continuous map. In fact, the conditions on F ensure that H is a homeomorphism. Next, $H(p, 0) = F(p, 1) = g(p)$. Thus H restricted to M_0 in W' coincides with g. It follows that the map HG is a homeomorphism of W on W' whose restriction to M_0 is gg^{-1} = identity. Summing up the result:

Lemma 6-1. *Let ϕ and ϕ' be modifications on M_0 satisfying the conditions described in the foregoing. Then ϕ and ϕ' have the same results and there is a homeomorphism between their traces whose restriction to M_0 is the identity.*

This can now be applied to sequences of modifications. Suppose, for example, that a sequence of two modifications ϕ_1 and ϕ_2 is applied to M_0, ϕ_1 transforming M_0 into M_1 and ϕ_2 transforming M_1 into M_2. Suppose also that ϕ_2 is replaced by a modification ϕ_2', related to ϕ_2 in the same manner as ϕ' and ϕ in Lemma 6-1. Write W_1, W_2, W_2' for the traces of ϕ_1, ϕ_2, ϕ_2', respectively. Then the trace of the sequence ϕ_1, ϕ_2 is $W = W_1 \cup W_2$ and that of the sequence ϕ_1, ϕ_2 is $W' = W_1 \cup W_2'$, in each case the union being formed with the identification of the points of M_1. Lemma 6-1 implies at once that there is a homeomorphism of W onto W'; in fact, it is the identity on W_1.

A similar idea can be applied to sequences of any number of modifications.

Exercise. **6-5.** Let S be a directly embedded sphere in M_0, with tubular neighborhood B expressed as a product $S \times E$. Let S' be a second sphere, a submanifold of B, meeting each cellular slice of B at one point, and let B' be a tubular neighborhood of S' contained in B and such that each slice of B' is contained in a slice of B. Prove that S, S', B, B' satisfy the conditions of Lemma 6-1.

The result of this exercise is going to be important in Section 6-8 on the rearrangement of modifications.

6-7. GENERAL POSITION

The idea on which the results of the next section are to be based is quite easy to grasp intuitively, but the details of the proofs are too complicated to be given here. The concept involved will be motivated by means of examples.

Consider first two curves in the plane, intersecting at a point p. If one or other of the curves is displaced slightly in the plane, the displaced curves will continue to have a point of intersection near p (Fig. 6-6). On the other hand, if one of the curves is displaced out of the plane into 3-space, the inter-

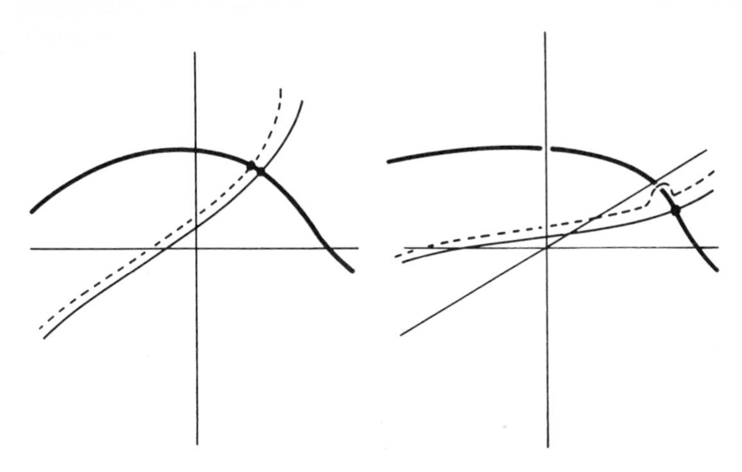

FIGURE 6-6 *Displaced blue curve in the plane still cuts the black curve. Displace it upwards into 3-space and the intersection is removed.*

section is removed. Of course, a pair of curves in the plane may not have isolated intersection points. But if, for example, two curves have an arc in common, a small displacement of one will give a pair of curves with isolated intersections (Fig. 6-7). The idea illustrated here is that a pair of curves in general position in the plane have isolated points of intersection, whereas if they are in general position in 3-space, they have no points in common. Also, intersections of curves in 3-space can be removed by small displacements.

To see what part the dimensions play in this we should think of points in the plane as having two degrees of freedom. That is, two coordinates have to be given to fix a point. A point on a curve, however, has just one degree of freedom, and so the condition of lying on a curve removes one degree of freedom. Thus the condition of lying on two curves removes two degrees of freedom. In general, therefore, the set of intersections of the two curves should have no degrees of freedom, and so should consist of isolated points. On the other hand, a point in 3-space has three degrees of freedom, whereas if it is to lie on a curve, it will have only one. Thus the condition of lying on a curve in 3-space removes two degrees of freedom. The condition of lying on two curves will remove four degrees of freedom, but since there are, in fact, only three degrees of freedom, two curves in 3-space should in general have no intersections at all. In addition, the discussion of Figs. 6-6 and 6-7 suggests that general position in 3-space can be achieved by means of small displacements.

Now, to carry out a similar intuitive argument in a more general context, let M be a differentiable manifold of dimension n and N a submanifold of dimension r. A point of M has n degrees of freedom, but if it lies on N, it has only r. The condition of lying on N thus removes $n - r$ degrees of freedom. If N' is a second submanifold of dimension r', then the condition of lying on both N and N' should in general remove $n - r + n - r'$ degrees of freedom. If this number turns out to be greater than n, we would expect no intersections. That is, submanifolds of dimensions r and r' with $r + r' < n$ should, if they are in general position, be disjoint. Moreover, given any submanifolds N and N', it should be possible to attain this general position by a slight displacement of one of them. Here a displacement of N, say, means

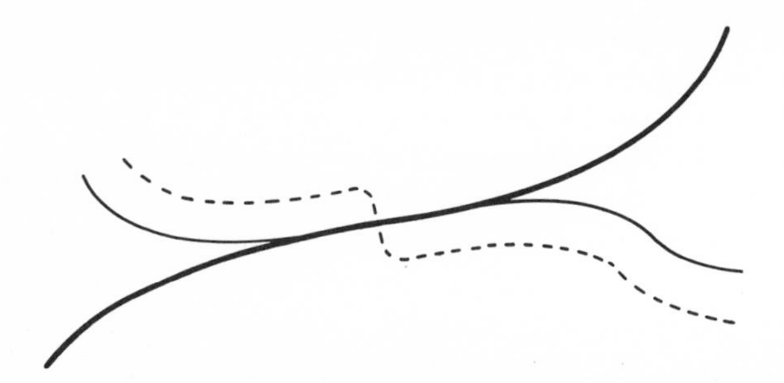

FIGURE 6-7 *Intersections of curves in the plane are made isolated by displacement.*

replacing the inclusion map $f: N \to M$ by another map $g: N \to M$ such that $g(N)$ is a submanifold and $g(p)$ is always near $f(p) = p$.

6-8. REARRANGEMENT OF MODIFICATIONS

The results of the last two sections will now be combined to prove an important theorem on sequences of modifications. The first step is to show that, under suitable conditions, the order of performance of two modifications can be switched around.

So let ϕ_1 be a modification of type r transforming a given compact differentiable manifold M_0 into M_1 with trace W_1 and let ϕ_2 be a modification of type s transforming M_1 into M_2 with trace W_2. ϕ_1 and ϕ_2 are to shrink spheres S^r in M_0 and S^s in M_1 to points P_1 and P_2 in W_1 and W_2, respectively, and introduce spheres S^{n-r-1} and S^{n-s-1}.

Now suppose that $s \leq r$. Then the discussion of Section 6-7 shows that a slight displacement of S^s can be made so that it does not meet S^{n-r-1}. By Exercise 6-5 this displacement of S^s can be done so that the conditions of Lemma 6-4 are satisfied, thus neither the result nor the trace of the sequence ϕ_1, ϕ_2 of modifications is affected. In addition, S^s and S^{n-r-1}, now being disjoint, have disjoint neighborhoods. Since a modification can be defined by using arbitrarily small tubular neighborhoods of the spheres to be shrunk (remark following Exercise 6-4), it follows that ϕ_2 and ϕ_1 in reverse can be defined by using disjoint tubular neighborhoods of S^{n-r-1} and S^s, and again this makes no difference to the result of the sequence ϕ_1, ϕ_2 or to its trace. The following has thus been proved.

Lemma 6-2. *Let ϕ_1 and ϕ_2 be as described at the beginning of this section. Then, without changing the result or the trace of this sequence of modifications, it can be assumed that the S^s in M_1 shrunk by ϕ_2 is disjoint from the S^{n-r-1} introduced by ϕ_1. In fact, the tubular neighborhoods of these spheres corresponding to the modifications ϕ_2 and ϕ_1 in reverse are disjoint.*

This lemma has an immediate consequence. For let B_1 and B_2 be the disjoint tubular neighborhoods of S^{n-r-1} and S^s mentioned there. These would be the sets removed from M_1 to start the construction of ϕ_1 in reverse and ϕ_2, respectively. Now, referring back to the construction of the trace of a modification, it will be seen that W_1 is the union of $(M_1 - \operatorname{Int} B_1) \times I$ with an $(n+1)$-cell E_1, while W_2 is the union of $(M_1 - \operatorname{Int} B_2) \times I$ with an $(n+1)$-cell E_2, in each case with the appropriate identifications. Also $E_1 \cap M_1$ is exactly B_1 and $E_2 \cap M_1$ is B_2. Now it is easy to see that W_1 contains the set $B_2 \times I$, as a subset of $(M_1 - \operatorname{Int} B_1) \times I$, while W_2 contains $B_1 \times I$, as a subset of $(M_1 - \operatorname{Int} B_2) \times I$. Also, the union of E_1 and $B_1 \times I$ is still an $(n+1)$-cell, E_1' meeting M_2 in a set B_1' homeomorphic to $S^{n-r-1} \times E^{r+1}$, and the union of E_2 and $B_2 \times I$ is an $(n+1)$-cell, E_2' meeting M_0 in a set B_2' homeomorphic to $S^s \times E^{n-s}$.

Exercise. 6-6. Check the statement made in the last sentence.

This all means, however, that the two modifications ϕ_1 and ϕ_2 can be thought of as performed simultaneously on M_0. That is, if the sets B_0 and B_2' are removed from M_0 and replaced by B_1' and a set $S^{n-s-1} \times E^{s+1}$, all with the appropriate identifications, M_2 is obtained. Moreover, the trace $W = W_1 \cup W_2$ of the pair of modifications ϕ_1, ϕ_2 is constructed by adding the cells E_1' and E_2' to $(M_0 - \operatorname{Int} B_0 - \operatorname{Int} B_2') \times I$ with the appropriate identifications. Hence, both with respect to the final result and to the construction of the trace, the two modifications ϕ_1 and ϕ_2 appear on an equal footing.

Note that all this has been done with the hypothesis $s \leq r$.

The point of this was to ensure that S^{n-r-1} and S^s in M_1 could be made disjoint. The same conclusion would hold if these spheres were given to be disjoint, regardless of the relation between s and r.

Continuing with the main argument, because ϕ_1 and ϕ_2 now appear on an equal footing, we can switch the order of performing them. That is, we can construct the modification ϕ_2 first and then perform ϕ_1 on the result, and the final result and the trace will be the same as for the given sequence ϕ_1, ϕ_2. This gives the main theorem of this section.

Theorem 6-4. *If M_2 is obtained from M_0 by two modifications, ϕ_1 of type r followed by ϕ_2 of type s with $s \leq r$, then the same result M_2 can be obtained by a modification of type s followed by a modification of type r, and the trace of the new sequence will be the same as that of the given one.*

Repeated application of this result yields the following theorem.

Theorem 6-5. *If a sequence of modifications is given, it can be rearranged, without changing final result or trace, so that modifications of type s are performed before those of type r whenever $s \leq r$.*

6-9. AN APPLICATION TO 3-MANIFOLDS

Let M be a compact orientable three-dimensional differentiable manifold. If two disjoint cells are removed from M, a manifold M' is obtained with a boundary consisting of two disjoint 2-spheres M_0 and M_2. Thus M' can be thought of as the trace of a sequence of modifications transforming M_0 into M_2. In this case the only possible types of modification are 0 and 1, and by Theorem 6-5 it can be assumed that all those of type 0 are done first, giving a manifold M_1. Then the modifications transforming M_1 into M_2 are all of type 1 or, what comes to the same thing, the modifications transforming M_2 into M_1 are all of type 0. Remember that all this rearrangement does not affect M'. Replacing the 3-cells that were removed, we now see that M is the union of two manifolds W_1 and W_2 with the common boundary M_1. Since the modifications leading from a 2-sphere to M_1 are all of type 0 and of the orientable kind, W_1 and W_2 are both solid spheres with handles. Hence, the following has been proved.

Theorem 6-6. *A compact orientable three-dimensional manifold is the union of two solid spheres with handles, the surfaces being identified.*

6-10. INTERPRETATION OF THEOREM 6-5 IN TERMS OF CRITICAL POINTS

Throughout this chapter not much attention has been paid to making the constructions differentiably. To do this requires some extra care. For example, when the trace of a modification is constructed, the various pieces must be put together to form a differentiable manifold. It was remarked that this can be done, and in Theorem 6-1 it was stated without proof that each modification of type r will then correspond to a differentiable function on the trace with a critical point of type number $r + 1$. Similarly, the construction described in Section 6-8 for rearranging modifications can be done in such a way that the pieces are always put together to form differentiable manifolds. Thus, suppose that M is a differentiable manifold with boundary $M_0 \cup M_1$, regarded as the trace of a sequence of modifications transforming M_0 into M_1, and suppose that the sequence is arranged as described in Theorem 6-5. It then follows that there is a differentiable function f on M, with values between 0 and 1, equal to 0 on M_0 and equal to 1 on M_1, with a finite number of nondegenerate critical points such that if P_1 and P_2 are two of them with the type number of P_1 less than that of P_2, then $f(P_1) < f(P_2)$. Such a function is described in [7] as a *nice function* on M.

7

Two-Dimensional Manifolds

7-1. INTRODUCTION

As an illustration of the use of the ideas introduced up to this point, the classification of two-dimensional manifolds will now be described. In the classical approach to this problem the manifolds are given as simplicial complexes that are then reduced to a set of canonical forms by a sequence of cutting and pasting operations. It is shown in this way that a compact connected orientable 2-manifold is homeomorphic to a sphere with a number of handles attached, and a compact connected nonorientable 2-manifold is homeomorphic to a sphere in which circular holes have been cut, after which diametrically opposite pairs of points on the circumferences of the holes have been identified. In each case the number of handles or number of holes is a topological invariant of the surface. These results will now be obtained by taking the manifolds to be differentiable and then by studying the critical points of functions on them. The orientable and nonorientable cases will be examined separately.

7-2. ORIENTABLE 2-MANIFOLDS

Let M be a compact, connected, orientable 2-manifold and, as in Theorem 4-2, construct on it a function f with a finite number of nondegenerate critical points. In this situation there are just three types of critical points, minimum, saddle point, and maximum, corresponding to type numbers 0, 1, and 2, respectively. The argument of Theorem 6-5 shows that f can be so chosen that the minima all correspond to smaller values of f than the saddle points and these in turn correspond to smaller values than the maxima. In Fig. 7-1, M is pictured as being in 3-space with f equal to the value of the last coordinate, and the critical points are arranged as just described. Incidentally, it is not yet obvious, although it happens to be true, that M can be embedded in a space of dimension three.

As in Chapter 6 we can now think of M as the trace of a sequence of modifications on one-dimensional manifolds. Each minimum corresponds to the introduction of a circle, so that,

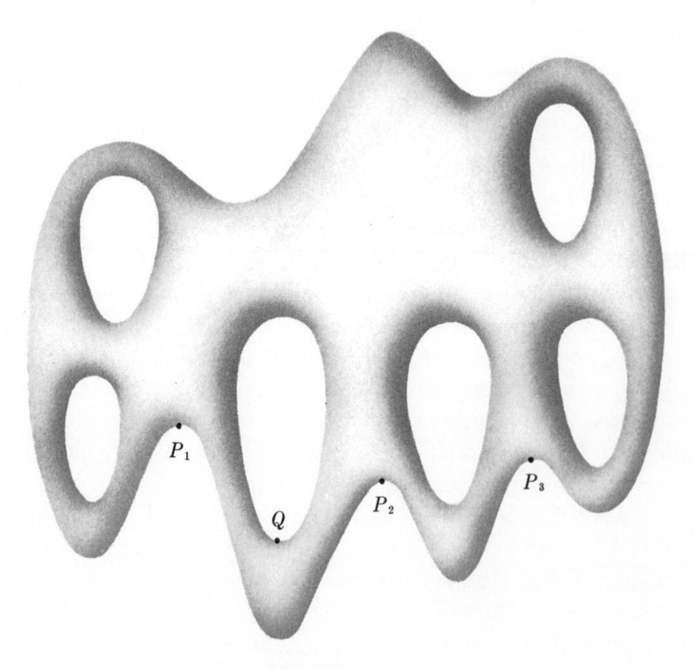

FIGURE 7-1

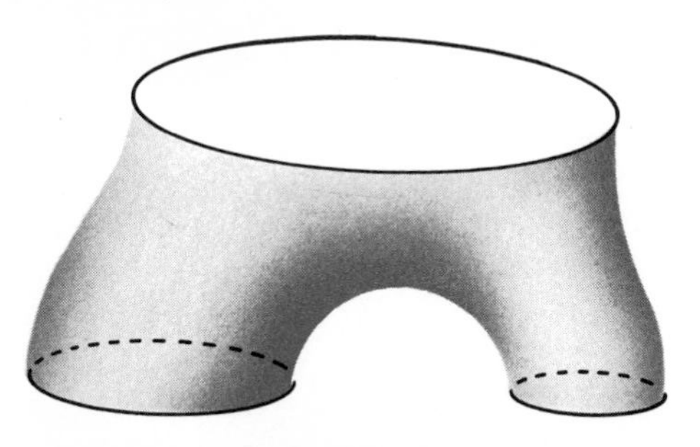

FIGURE 7-2

if we consider the family of levels of f, starting from the bottom, a finite number of circles will have been introduced after all the minima are passed. Next, on this 1-manifold, a union of circles, a number of modifications of type 0 are performed, corresponding to the saddle points on M. Finally, modifications are performed corresponding to the various maxima of f, a circle being extinguished at each such point.

Considering first the type 0 modifications, it will be noticed that these are of three kinds, which can be conveniently called connecting, disconnecting, and twisting. A connecting modification performed on a level M_c of f reduces the number of its components by one, two of the circles being joined to form one. Figure 7-2 shows part of M that forms the trace of such a modification. For example, in Fig. 7-1 the modification corresponding to the critical point P_1 is of the connecting kind. The disconnecting kind works in exactly the opposite way (cf. Fig. 7-3). In Fig. 7-1, for example, the modification corresponding to Q is a disconnecting modification.

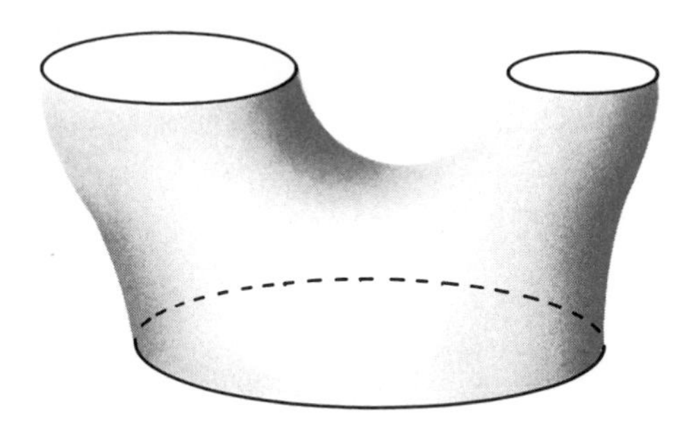

FIGURE 7-3

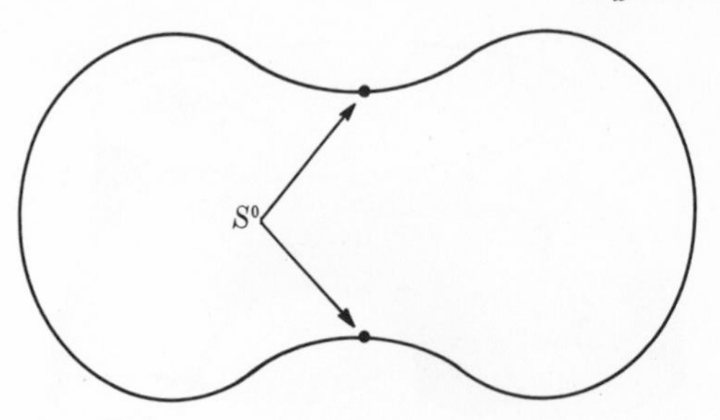

FIGURE 7-4

In the third kind of modification of type 0, a 0-sphere on a circle is to be shrunk (cf. Fig. 7-4), but this time the circle is twisted over in a neighborhood of one of the component points of the 0-sphere (Fig. 7-5). The result of the modification (Fig. 7-6) is again a circle. The twisting of the circle in this kind of modification means that the level curves of the corresponding function are not plane curves, and so the trace of the modification cannot be contained nonsingularly in Euclidean 3-space.

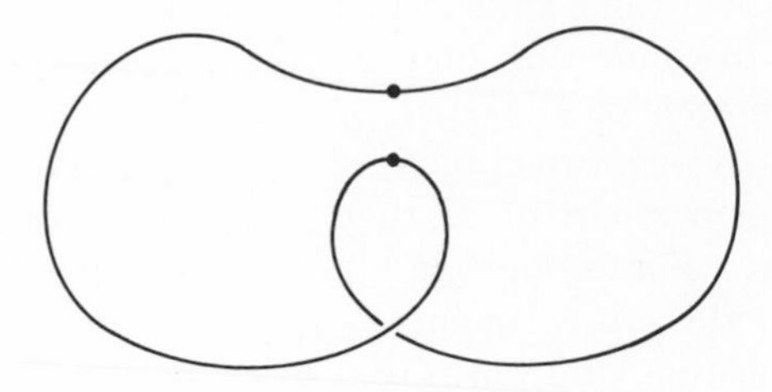

FIGURE 7-5

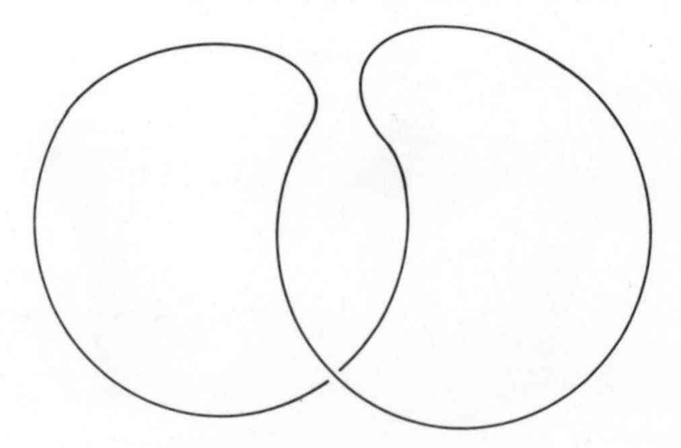

FIGURE 7-6

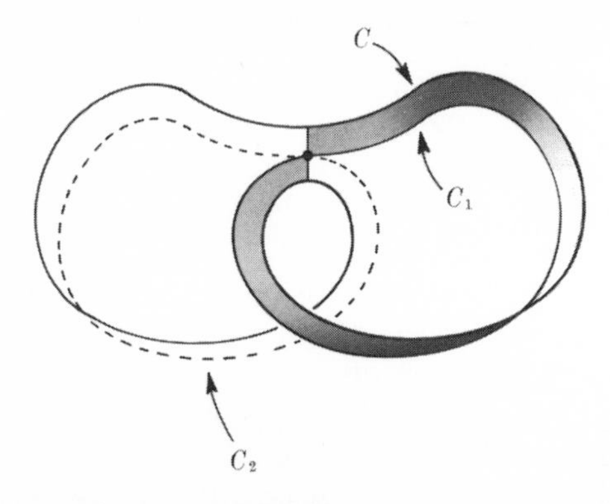

FIGURE 7-7

Consider now the trace W of the modification illustrated in Figs. 7-4, 7-5, and 7-6. There is a function f on W having one critical point P. The corresponding critical level consists of two circles C_1 and C_2 intersecting at P (cf. Fig. 7-7). The third circle C in Fig. 7-7 is supposed to be a noncritical level of f below P; in other words, a circle on which the twisting modification is to be performed. The shaded strip in Fig. 7-7 is supposed to be the part of W between C and C_1. Also, Fig. 7-8 represents a rectangle with the ends identified as shown by the arrows to form a Möbius strip. The shaded strip on W in Fig. 7-7 can be mapped on the shaded part of the Möbius strip by identifying C_1 with the center line of the strip and identifying arcs on W orthogonal to the levels of f with the vertical segments on the rectangle of Fig. 7-7. Similarly, the other half of the Möbius strip can be identified with part of W between C_1 and a noncritical level above P. Thus C_1 has a neighborhood on W that is a Möbius strip. That is, C_1 is not directly embedded in W, which is therefore nonorientable. For the moment, orientable manifolds only are being discussed, and so the 0-type modifications will be all connecting or disconnecting.

Suppose now that $C_1 \cup C_2 \cup \cdots \cup C_n$ is a level M of f above all the minima. Thus it is the 1-manifold to which the

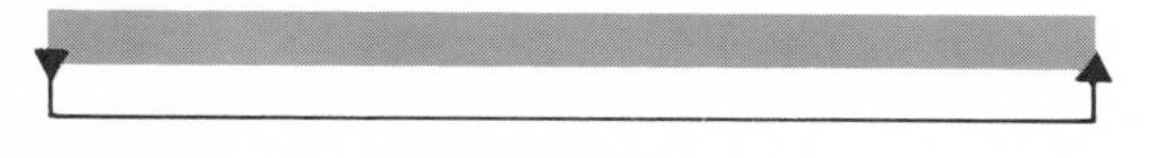

FIGURE 7-8

type 0 modifications are to be applied. And suppose that the type 0 modifications can be divided into two sets, one operating only on $C_1 \cup C_2 \cup \cdots \cup C_{n-1}$ and the other only on C_n. The traces of these two sets, with cells added at the top and bottom corresponding to the maxima and minima of f, would then be disjoint 2-manifolds whose union would be M, contradicting the connectedness of M. Hence, there must be a modification, necessarily of the connecting kind, shrinking a 0-sphere that consists of a point on C_n and a point on one of the other C_i. Rearrange the 0-type modifications so that this one is done first. The remaining modifications of type 0 then operate on a union of $n - 1$ circles. Repeating this argument, it follows that the given sequence of modifications can be rearranged so that the first $n - 1$ of them join the n circles C_i to form a single circle C_0. Geometrically this means that the saddle points P_i in Fig. 7-1 are pulled down so that they lie below the others.

Now any two of the C_i, say C_1 and C_2, are the boundaries of 2-cells E_1 and E_2 on M. If they are joined by a connecting modification, the resulting circle also bounds a 2-cell (Fig. 7-9) obtained by adding E_1 and E_2 to the trace of the modification.

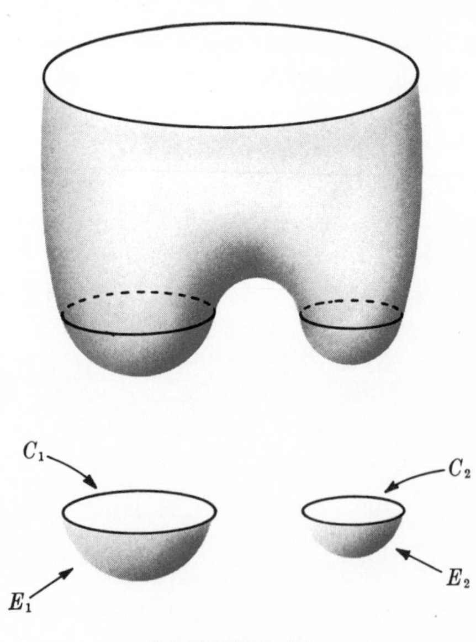

FIGURE 7-9

Repeating this $n - 1$ times, we find that the part of M below C_0 is a 2-cell E_0. We can now replace f by a new function whose level curves above C_0 are as before, but whose level curves below C_0 form a family of circles shrinking to a point on E_0. That is, f will have just one minimum on M.

A similar argument shows that the initial function f can be adjusted so that it has just one maximum on M. The result so far obtained can be formulated as follows.

Lemma 7-1. *On a compact connected orientable 2-manifold there is a function whose critical points are all saddle points except for one minimum and one maximum.*

Alternatively this lemma says that the manifold can be regarded as the trace of a sequence of modifications of type 0, starting with a circle C_0 and ending with a circle C_1, with cells E_0 and E_1 added at the top and bottom (Fig. 7-10). With this terminology, the next step is to rearrange the modifications leading from C_0 to C_1 so that all the disconnecting modifications are done first.

Now the cell E_0 (cf. Fig. 7-10) is homeomorphic to a 2-sphere with a hole in the surface, the hole having boundary C_0. If the trace of a disconnecting modification (cf. Fig. 7-3) is added, with identification along C_0, the result is a sphere with two holes in the surface. Proceeding inductively, we see that, if the trace of $k - 1$ disconnecting modifications is added to E, the result is a sphere M_1 with k holes in it having boundaries $\Gamma_1, \Gamma_2, \ldots, \Gamma_k$, say. The remainder M_2 of M is obtained as the trace of connecting modifications on the union of the Γ_i, the cell E_1 being added on top. However, connecting modifications leading from C_0 to C_1 are disconnecting modifications leading from C_1 to C_0, so that M_2 is also a sphere with k holes. M is then constructed by forming the union of M_1 and M_2 with identification of the boundaries of the holes in pairs. The following result has thus been proved.

Lemma 7-2. *A compact connected orientable 2-manifold is homeomorphic to a manifold $M(k)$, for some k, where $M(k)$ is the union of two spheres each with k holes, the circumferences of the holes being identified in pairs (cf. Fig. 7-11).*

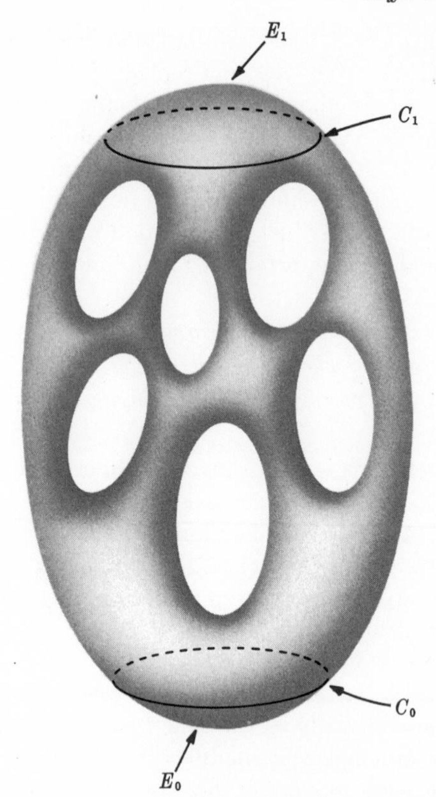

FIGURE 7-10

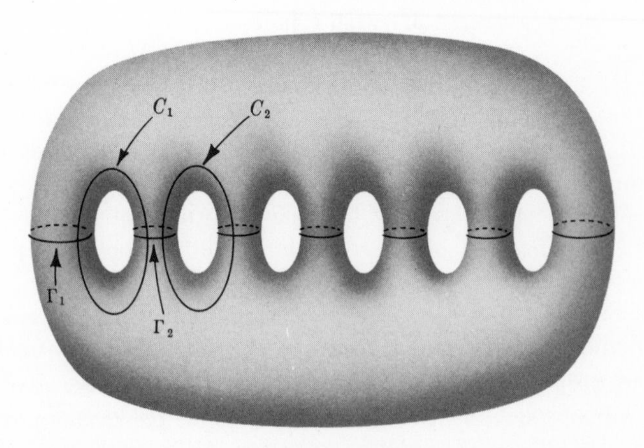

FIGURE 7-11

This lemma yields a sequence of manifolds, namely the $M(k)$, such that every compact, connected orientable 2-manifold is homeomorphic to one of them. But to give this real meaning as a classification theorem, it will have to be shown that, for $h \neq k$, $M(h)$ and $M(k)$ are not homeomorphic. However, the $M(k)$ are not in the form normally used for classifying the 2-manifolds, and so first Lemma 7-2 will be translated into a more usual form.

To make the translation, $M(k)$ will be looked at from a different point of view: suppose a set of $k - 1$ circles $C_1, C_2, \ldots, C_{k-1}$ is drawn on $M(k)$ as indicated in Fig. 7-11. If $M(k)$ is split into two spheres each with k holes with boundaries $\Gamma_1, \Gamma_2, \ldots, \Gamma_k$, then on each copy a set of arcs is drawn joining a point of Γ_i to a point of Γ_{i+1} for each i, and the C_i are obtained by joining up these arcs in pairs to form circles. The C_i are directly embedded in $M(k)$, and if the modifications on $M(k)$ that shrink these circles are performed, the result is a 2-sphere. Reading this statement backward, we see that $M(k)$ is the result of performing $k - 1$ modifications of type on the 2-sphere, all the modifications being of the orientable kind (cf. Example 6-10). Each such modification has the effect of attaching a handle to the 2-sphere (Fig. 6-2). Thus if Σ_{k-1} denotes the surface of a 2-sphere with $k - 1$ handles, it follows that $M(k)$ is homeomorphic to Σ_{k-1}.

With a suitable change of notation, Lemma 7-3 takes the following form.

Lemma 7-3. *A compact connected orientable 2-manifold is homeomorphic to Σ_p for some p, where Σ_p is a sphere with p handles.*

Exercise. **7-1.** Give an alternative proof that $M(k)$ is homeomorphic to Σ_{k-1} by applying to Σ_{k-1} the sequence of steps in the proofs of Lemmas 7-1, 7-2.

The final step in this discussion is to show that Σ_p and Σ_q are not homeomorphic if $p \neq q$.

First note that there are p disjoint circles on Σ_p, one for each handle, as indicated in Fig. 7-12, such that if Σ_p is cut along them, the resulting surface is still connected. The

FIGURE 7-12

maximum number of disjoint circles with this property on a surface is clearly a topological invariant of the surface.

Definition 7-1. The maximum number of disjoint circles along which a surface can be cut without disconnecting it is called the *genus* of the surface.

There is a theorem, which will not be proved here, that says that the genus of a compact 2-manifold is finite, and in particular that of a sphere is zero (cf. [9]). This theorem will be needed in the following discussion. The proof that Σ_p and Σ_q are not homeomorphic when $p \neq q$ is essentially a matter of proving that the genus of Σ_p is p.

Lemma 7-4. *Let the surface M' be obtained from M by a modification of type* 0. *Then the genus of M' is greater than that of M.*

Proof. Let the genus of M be p and let $C_1, C_2, \ldots, C_p$ be disjoint circles on M such that M remains connected when cut along them. It can be assumed (cf. Lemma 6-1) that the 0-sphere to be shrunk in transforming M to M' does not meet the C_i. Then (Fig. 7-13) $C_1, C_2, \ldots, C_p$ still appear in M', but in addition there is a circle C_{p+1} around the added handle,

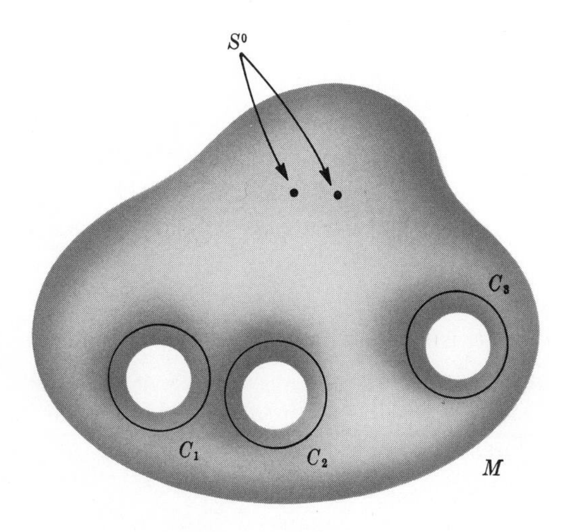

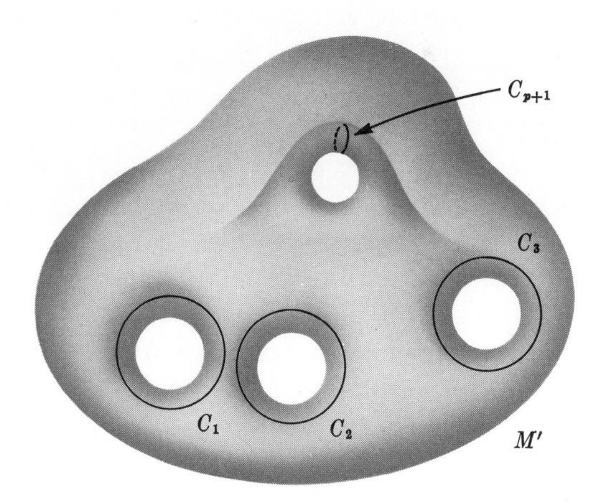

FIGURE 7-13

not meeting $C_1, C_2, \ldots, C_p$ and such that M' can be cut along $C_1, C_2, \ldots, C_{p+1}$ without disconnecting it. Hence, the genus of M' is at least $p + 1$, which is greater than the genus of M.

If this lemma is applied repeatedly, the following result is obtained.

Lemma 7-5. *Let the genus of M be p and let M' be obtained from M by p modifications of type* 1, *remaining connected all the time. Then M' is a sphere.*

Proof. After each modification in the transformation from M to M', Lemma 7-4 implies that the genus is decreased by at least 1. It follows that after p modifications, the genus must be zero. But the resulting M' will be a sphere with q handles for some q, and since the genus is zero, q must be zero. Thus M' is a sphere.

Note. The modifications leading from M to M' here must, in fact, shrink a set of p disjoint circles on M. Also, looking at the process backward, this means that the surface M of genus p must be homeomorphic to a sphere with p handles. The essential content of Lemma 7-5 for the present purpose is that a finite number (equal to the genus) of modifications of type 1 will reduce an orientable surface to a sphere, and any further modifications will disconnect it, since the genus of a sphere is 0.

The main result of the section can now be proved.

Theorem 7-1. *A compact connected orientable* 2*-manifold is homeomorphic to some Σ_p, and any two Σ_p and Σ_q with $p \neq q$ are not homeomorphic.*

Proof. The first part was proved in Lemma 7-3. Suppose now that Σ_p is homeomorphic to Σ_q with $q > p$. A sequence of $q - p$ modifications of type 1 can be performed on Σ_q, shrinking circles round $q - p$ of its handles, giving Σ_p as the result. But the assumption that Σ_p and Σ_q are homeomorphic means that these modifications can be thought of as performed on Σ_p, giving Σ_p as the result. Applying the same modifications repeatedly would result in an arbitrarily long sequence of modifications of type 1 operating on Σ_p and leaving it connected. This contradicts the earlier remark that after a certain finite number of modifications of type 1 on a compact orientable 2-manifold it must become disconnected. Hence, Σ_p and Σ_q cannot be homeomorphic.

Exercises. **7-2.** Prove that if a modification of type 1 is performed on Σ_p, the result, if connected, is Σ_q with $q < p$.

(*Hint:* Show that if $q \geq p$, there would be an arbitrarily long sequence of modifications of type 1 on Σ_p leaving it connected.)

7-3. Prove that the genus of Σ_p is p. (It is trivial that the genus of Σ_p is at least p. Suppose it is greater. Construct a sequence of modifications, using the last exercise, that will reduce Σ_p to a sphere, but will leave it with genus greater than 0.)

7-3. THE NONORIENTABLE CASE

Before looking at the nonorientable case in general, it will be helpful to examine a simple special case, to get some understanding of the way in which a twisting modification works (cf. Figs. 7-4, 7-5, 7-6). Consider, then, the projective plane; it can be represented as the surface of a 2-sphere with all pairs of diametrically opposite points identified. Thus if the sphere is taken to be $x^2 + y^2 + z^2 = 1$ in 3-space, the point (x, y, z) is to be identified with $(-x, -y, -z)$. Since the function

$$f(x, y, z) = x^2 + 2y^2 + 3z^2$$

takes the same value at $(-x, -y, -z)$ as at (x, y, z), this formula defines a function f on the projective plane. We can now work out the behavior of the level curves of f on the projective plane by examining the levels of $x^2 + 2y^2 + 3z^2$ on the sphere, remembering that pairs of opposite points are to be identified.

The level curves on the sphere are the intersections of the sphere with the family of ellipsoids

$$x^2 + 2y^2 + 3z^2 = c.$$

Clearly, if $c < 1$, there are no real intersections. When $c = 1$, there are two points of intersection, namely $(\pm 1, 0, 0)$ (see Fig. 7-14). This, of course, means one point on the projective plane, and will correspond to the minimum P_0 of f on the projective plane. As c increases from 1 to 2, the intersection will be a pair of ovals on the sphere (Fig. 7-15). The identification of opposite points means that this appears as one circle on the projective plane. When $c = 2$, the intersection of the ellipsoid and the sphere is as in Fig. 7-16. Thus f has a critical point at $(0, \pm 1, 0)$, which again represents one point P_1 on the pro-

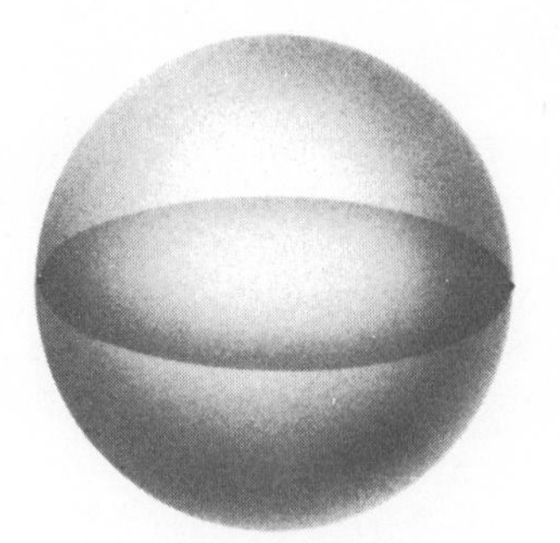

FIGURE 7-14

jective plane. Then as c varies from 2 to 3 the intersection of the sphere and the ellipsoid is again a pair of circles, as in Fig. 7-17, representing one circle on the projective plane. These shrink, as c tends to 3, to the points $(0, 0, \pm 1)$ on the sphere, representing one point P_2 on the projective plane.

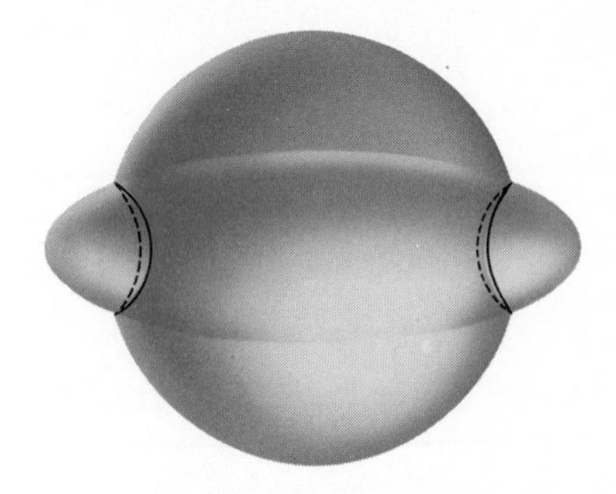

FIGURE 7-15

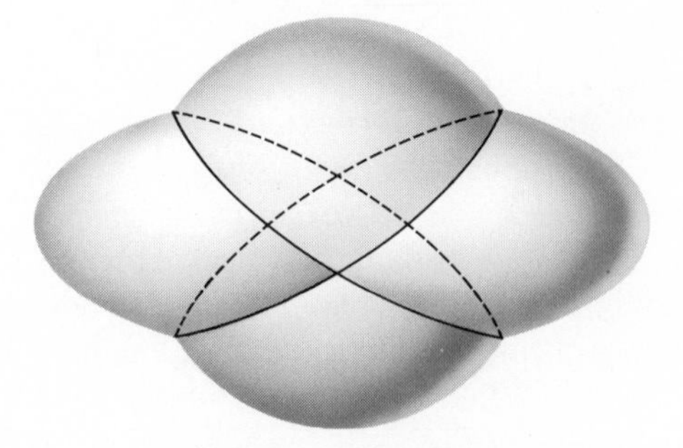

FIGURE 7-16

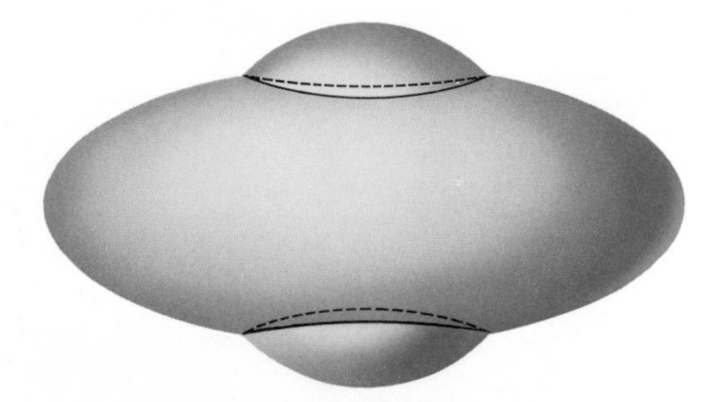

FIGURE 7-17

Examining more closely the transition across the critical level through P_1, we note that the projective plane can be regarded as the hemisphere $y \leq 0$ on the unit sphere, with the opposite points of the circumference identified. For convenience this hemisphere is flattened out into a disk in Fig. 7-18. This figure shows the critical level of f through P_1, as well as noncritical levels on either side of it. The arcs a and b form the lower noncritical level, the arcs c and d the upper one, and the lettering round the edge indicates the identification of opposite points. The arrows on a and b show a direction of rotation round the lower level of f, and it will be noticed that in a neighborhood of P_1 they lie in the same direction. Comparing this figure with Fig. 7-5 permits us to see that the transition from the lower to the upper noncritical level involves a twisting modification.

Another way of describing the foregoing situation is that the trace of a twisting modification is a projective plane with

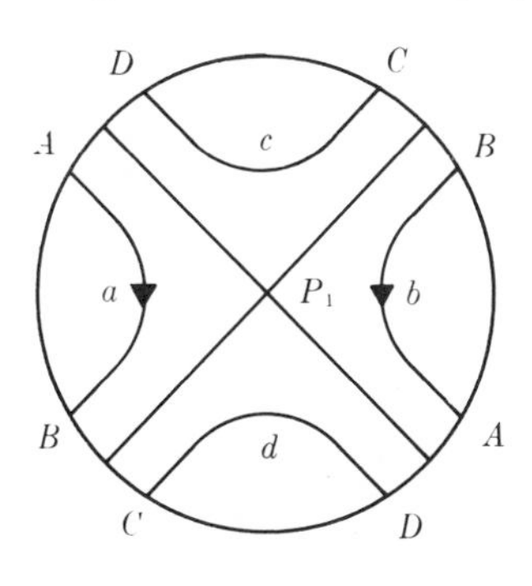

FIGURE 7-18

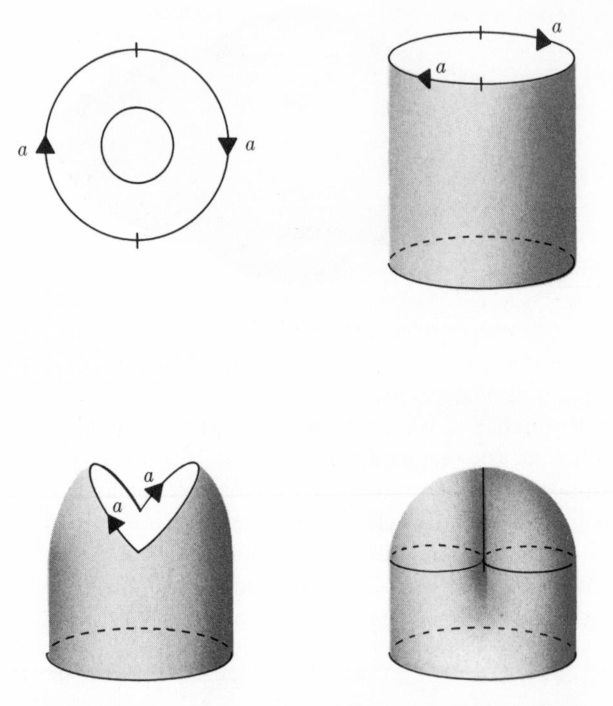

FIGURE 7-19 Step-by-step construction of a crosscap.

two holes cut in it. Thus the trace of a sequence of k twisting modifications can be obtained by taking k projective planes, each with two holes in it, and identifying the boundary of one hole in each plane with the boundary of one hole in the next.

There is another convenient terminology that can be used in the last construction. Let M_1 and M_2 be any two connected manifolds of the same dimension. Take points P_1 and P_2 on M_1 and M_2, respectively, and remove cellular neighborhoods U_1 and U_2 of P_1 and P_2, respectively. Form the union of $M_1 - U_1$ and $M_2 - U_2$, identifying the boundaries of U_1 and U_2. The result is called the *connected sum of M_1 and M_2*. Note that it is the result of a modification of type 0 performed on $M_1 \cup M_2$, shrinking the sphere $P_1 \cup P_2$.

Thus, using the terminology just introduced, the trace of a sequence of k twisting modifications is the connected sum of k projective planes with two holes cut in the final result.

The construction of the connected sum of a manifold M with a projective plane can be described alternatively as follows. The plane can be represented as a disk with diametrically opposite points of the circumference identified. To connect this to M, holes are cut in both M and the disk. The disk thus becomes an annulus or, what is the same thing, a cylinder, with the diametrically opposite points on one end identified. If this identification is actually carried out in 3-space (cf. Fig. 7-19), a surface called a *cross-cap* is obtained. Note that the self-intersection is simply the accidental result of trying to construct the cross-cap in 3-space. So the connected sum of M and the projective plane is obtained by identifying the boundary of the cross-cap with the boundary of a circular hole in M. This operation is called attaching a cross-cap to M. Note that the same result would be obtained by cutting a hole in M and identifying diametrically opposite points of its boundary.

The rest of the discussion of nonorientable 2-manifolds will be carried out in the following sequence of exercises.

Exercises. **7-4.** Let M be a compact connected nonorientable 2-manifold. Show, as in Section 7-2, that there is a function f on M with one maximum, one minimum, and a finite number of saddle points. Rearrange the saddle points so that the corresponding modifications of type 0 are arranged with the disconnecting kind done first, then the connecting kind, and finally the twisting kind. Hence, show that M is the connected sum of an $M(k)$, as in Lemma 7-2, and a number of projective planes. Or, in other words, M can be represented as a sphere with holes in it, some pairs of holes being joined by handles, other holes being filled by cross-caps.

7-5. In Fig. 7-20 the letters and arrows show identifications, so that the union of the two annuli with the given identifications is the connected sum of two projective planes, or two cross-caps joined base to

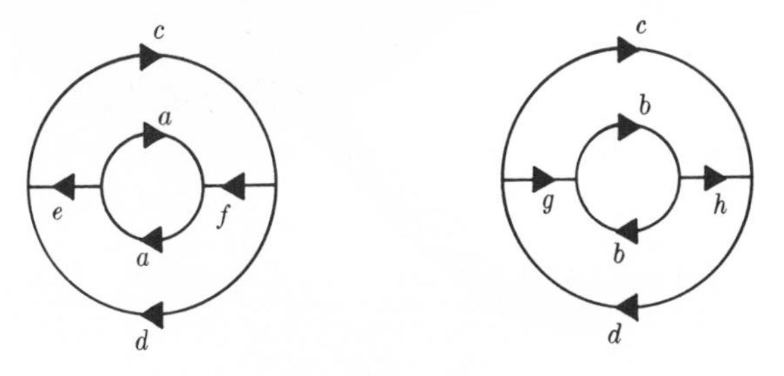

FIGURE 7-20

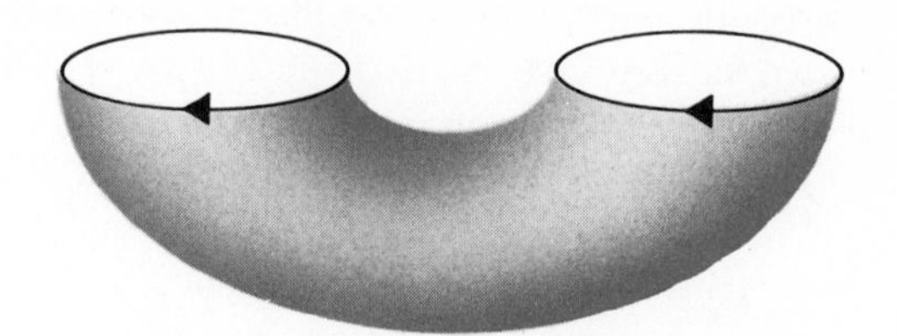

FIGURE 7-21

base. By cutting along e, f, g, h as shown and reassembling, show that this surface is homeomorphic to a cylinder with the ends identified as shown in Fig. 7-21. (This surface is the Klein bottle.) Figure 7-22 shows the identification carried out. Again note that there is a self-intersection when the surface is constructed in 3-space.

7-6. Give an alternative proof of the last exercise by constructing a function on the Klein bottle with one maximum, one minimum, and two saddle points corresponding to twisting modifications.

7-7. Note that the Klein bottle is the result of performing a modification of type 0 on the sphere in the nonorientable way (cf. Example 6-10).

Now if M is a projective plane and a type 0 modification is performed on it, there is no way of telling whether it is of the orientable kind or not. Hence, show that the connected sum of a Klein bottle and a projective plane is homeomorphic to the connected sum of a torus and a projective plane. Deduce that the connected sum of a torus and a projective plane is homeomorphic to the connected sum of three projective planes.

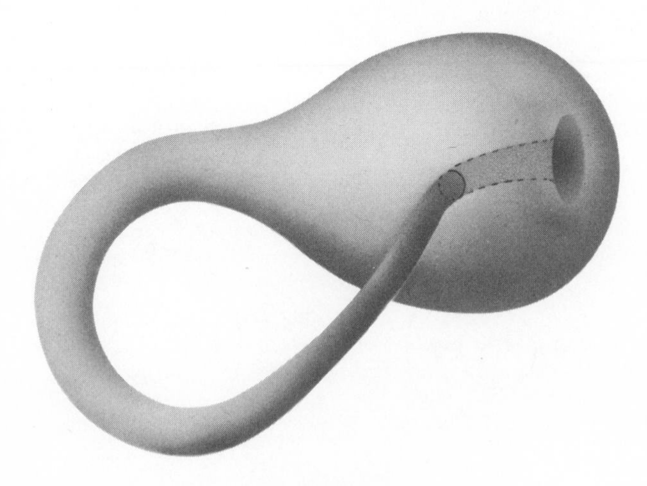

FIGURE 7-22

7-8. By repeated application of the last exercise, show that the connected sum of a Σ_p with any number of projective planes is itself a connected sum of projective planes.

7-9. The preceding exercises show that a compact connected nonorientable 2-manifold is homeomorphic to a manifold $N(k)$, the connected sum of k projective planes, for some k, or the result of attaching k crosscaps to a sphere. It is now to be shown that $N(h)$ is not homeomorphic to $N(k)$ for $h \neq k$. Check first that if N' is obtained from N by attaching a crosscap, the genus of N' is greater than that of N. Hence (cf. Lemma 7-5), show that there is a finite number of operations, each of which is either a type 1 modification or the removal of a cross-cap, that reduces a given surface to a sphere, while any further such operations will disconnect it.

7-10. Use an argument like that of Theorem 7-1 to show that $N(h)$ and $N(k)$ are not homeomorphic if $h \neq k$.

7-11. Prove that the genus of $N(k)$ is k.

8

Second Steps

The aim of this chapter is to give some indication of the way in which this subject can be developed beyond the very elementary ideas described up to now. To do this in detail would require a deeper knowledge of a number of topics in algebraic topology, and as this is beyond the scope of this book, the ideas treated here will only be sketched in an intuitive manner. To follow this up in more detail, the guide to further reading should be consulted.

8-1. KILLING OF HOMOTOPY CLASSES

The idea involved here has already been illustrated in Example 6-5, where a torus was transformed into a sphere by a modification of type 1. This transformation is a process of simplification. That is, the sphere can be thought of as simpler than the torus, in the sense that any closed path on the surface of the sphere can be shrunk to a point on the surface, whereas, for example, the circle a in Fig. 8-1 cannot be shrunk to a point on the surface of the torus. (These statements are

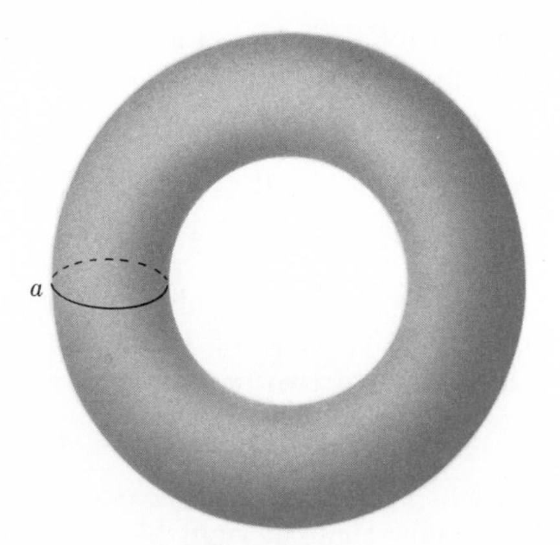

FIGURE 8-1

made as being intuitively obvious; for a rigorous treatment they need accurate formulation and proof.)

Consider now how this idea can be generalized. The first step is to describe a classification of closed curves in a manifold M. A closed path based on a point x of M is a continuous map $f\colon I \to M$ where I is the unit interval of real numbers, with the condition that $f(0) = f(1) = x$. Alternatively, f can be thought of as a map of a circle into M, a selected point of the circle being mapped on x. Two such paths f and g will be called homotopic, the notation being $f \sim g$, if there is a continuous map $F\colon I \times I \to M$ such that

$$\left.\begin{aligned} F(s, 0) &= f(s) \\ F(s, 1) &= g(s) \end{aligned}\right\}, \qquad \text{for all} \quad s \in I,$$
$$F(0, t) = F(1, t) = x, \qquad \text{for all} \quad t \in I.$$

Geometrically this means that F maps a square into M so that the bottom is mapped by f, the top by g, and the vertical sides onto x. And intuitively this means that if t is thought of as standing for time, then throughout a unit time interval the path f is continuously deformed into g. It turns out that this relation between paths is an equivalence relation, and so the set of closed paths based on x can be divided into equivalence classes, known in this context as homotopy classes.

The set of homotopy classes of closed paths on M based on x will be denoted by $\pi_1(M, x)$. This set can now be given an algebraic structure as follows. If f and g are two closed paths based on x, define fg to be the path obtained by tracing out f first, followed by g. The defining formulas for $h = fg$ are

$$h(s) = f(2s), \qquad 0 \leq s \leq \tfrac{1}{2},$$
$$h(s) = g(2s - 1), \qquad \tfrac{1}{2} \leq s \leq 1.$$

Then if $f \sim f'$ and $g \sim g'$, it can be shown that $fg \sim f'g'$ and so, denoting the homotopy class of f by $\bar{f}$, and similarly for other paths, the product of two homotopy classes can be defined by

$$\bar{f}\,\bar{g} = \overline{fg},$$

the point to notice being that the right-hand side depends only on the classes $\bar{f}$ and $\bar{g}$ and not on the particular paths f and g representing them. It can be shown (cf. [9]) that this multiplication is a group operation, the identity being the class of the constant path, which maps all of I on x, and the inverse being obtained by reversing paths. Thus $\pi_1(M, x)$ becomes a group, the fundamental group of M.

All the foregoing can be done for any topological space. However, since M is a compact differentiable manifold, it can be shown that $\pi_1(M, x)$ is finitely generated. Moreover, if the dimension of M is greater than 2, a general position argument (Section 6-7) shows that a given homotopy class always contains a path f that is a differentiable homeomorphism of a circle into M. If M is orientable, then this circle will be directly embedded (Definition 6-2).

Suppose now that M is an orientable differentiable manifold of dimension greater than 2, and let $\bar{\alpha}_1, \bar{\alpha}_2, \ldots, \bar{\alpha}_r$ be generators of $\pi_1(M, x)$. As described in the foregoing, $\bar{\alpha}_r$ can be represented by a circle α_r directly embedded in M. Perform a modification of type 1, shrinking α_r and transforming M into M'. Roughly speaking, this modification has the effect of supplying a disk (in M') of which α_r is the boundary. And so it turns out that in M' α_r is homotopic to a constant. In other words, the homotopy class $\bar{\alpha}_r$ becomes the identity. It can be shown that the transformation of M into M' does not, however, interfere with the other generators $\bar{\alpha}_1, \bar{\alpha}_2, \ldots, \bar{\alpha}_{r-1}$. Thus the fundamental group of M' is obtained from that of M by replacing the one generator $\bar{\alpha}_r$ by the identity. This operation is usually called *killing the class* $\bar{\alpha}_r$. Clearly, if this kind of operation is repeated for the other generators, the whole fundamental group can be killed.

This result can be stated by saying that a given orientable differentiable manifold can be transformed by spherical modifications of type 1 into a manifold with trivial fundamental group. Or, using Theorem 6-3, the given manifold cobounds a manifold with trivial fundamental group. Note that the condition that the dimension be greater than 2 is not needed, since an orientable 2-manifold is a sphere with handles, and it is easy to see that this can be transformed into a sphere by modifications of type 1, each shrinking a circle around a handle.

The ideas just presented can be generalized in yet another way. An element of $\pi_1(M, x)$ can be thought of as a homotopy class of circles in M. Other groups $\pi_r(M, x)$ can be defined whose elements are homotopy classes of r-spheres in M. Then classes represented by directly embedded r-spheres can be killed by modifications of type r. Operations of this type can lead to simplifications of the structure of the given manifold and can be of assistance in classification problems.

8-2. COMPLEMENTARY MODIFICATIONS AND CANCELLATION

It has already been seen that if M' is obtained from M by a modification of type r, shrinking S^r and introducing S^{n-r-1}, then we can return from M' to M by a modification of type $n - r - 1$, shrinking S^{n-r-1} and introducing S^r. Under certain conditions, however, there is a more interesting and less obvious way of reversing the effect of a modification. Consider first the following example.

Example

8-1. Start with a 2-sphere M and transform it into a torus by a modification ϕ of type 0 (orientable!) shrinking a 0-sphere S^0, as described in Example 6-4. Let B be a tubular neighborhood of S^0, namely, a pair of disjoint disks. The boundary of B is of the form $S^0 \times S^1$ (a pair of circles) and this contains a set $S^0 \times \{p\}$ for a fixed p, which can be thought of as a displaced copy of S^0.

Now S^0 is the boundary of a 1-cell E^1 in M. This can be adjusted so that it meets the boundary of B in $S^0 \times \{p\}$, and, to save extra notation, the part of E_0^1 in B will be dropped, so that $S^0 \times \{p\}$ appears as its boundary.

Now to get the torus M' the handle $E^1 \times S^1$ is attached to $M - B$. There is a segment $E^1 \times \{p\}$ on this handle whose end points join up with those of E_0^1 to form a circle S^1 on the torus. Clearly, if a modification ϕ' of type 1 is performed on the torus, shrinking this circle, the 2-sphere is recovered.

Examining this process more closely, we can obtain some more information. It has already been seen that the trace of ϕ is a solid torus with a spherical hole cut out of it, the boundary of the hole being M and the outer boundary of the torus being M' (Example 6-7). The level surfaces of the associated function (cf. Theorem 6-7) start off with M and are all 2-spheres until the critical level is reached. This level has the form of a torus with a circle pinched to a point. Beyond the critical level all the level surfaces are tori. It will be seen that, as we go through the set of level surfaces, the segment E appears on each one, until its ends are joined together on the critical level to form S^1, and then beyond that a copy of S^1 appears on each level surface. Now construct the combined trace of ϕ followed by ϕ' or, what comes to the same thing, attach the trace of ϕ' to that of ϕ that has already been constructed. The level surfaces in the trace of ϕ' can be thought of as an expanding family of tori with the circle S^1 shrinking to the critical point corresponding to the modification ϕ'. Beyond that critical level, the levels are all 2-spheres, and so the combined trace turns out to be a solid sphere with a spherical hole cut out of it. The interesting point to note is that this combined trace is actually of the form $M \times I$. In other words, the two modifications ϕ and ϕ' not only cancel, in the sense that the sequence ϕ followed by ϕ' leads back to M, but they also have the most trivial possible trace, namely, $M \times I$.

It will be seen that the essential point that makes the preceding example work is that the sphere S^r shrunk in the first modification is the boundary of a cell E_0^{r+1} in M, which is closed up to form a directly embedded sphere S^{r+1} in M'. Then the second modification ϕ' shrinks that sphere. It will now be shown that whenever this condition holds, the sequence of modifications ϕ followed by ϕ' has the trace $M \times I$.

So let S^r be a directly embedded sphere in a manifold M and suppose that S^r is the boundary of a cell E_0^{r+1} homeomorphically and differentiably embedded in M. A modification ϕ is going to be constructed, shrinking the sphere S^r with a product structure on a tubular neighborhood satisfying this additional condition: the tubular neighborhood B has the form $S^r \times E^{n-r}$ and its boundary has the form $S^r \times S^{n-r-1}$. Assume that E_0^{r+1} can be adjusted by a small displacement so that its intersection with the boundary of B is of the form $S^r \times \{p\}$, where p is a point of S^{n-r-1}. It will now be convenient to think of E_0^{r+1} as being contained in $M - B$; then its boundary sphere $S^r \times \{p\}$ is on the common boundary of B and $M - B$.

The construction of M', the transform of M by ϕ, is carried out by adding $E^{r+1} \times S^{n-r-1}$ to $M - B$ with suitable identification of boundary points. In particular, the cells $E^{r+1} \times \{p\}$ and E_0^{r+1} are joined along their boundaries to form a sphere S^{r+1} in M'. Assume the additional condition that this sphere S^{r+1} is directly embedded. There are ways of ensuring this by conditions on the product structure on B used in constructing ϕ, but this will not be discussed here. Of course, this condition is not automatically satisfied. It is not satisfied, for example, in the case of a modification of type 0 that is nonorientable.

Assuming, then, that S^{r+1} is directly embedded, let B' be a tubular neighborhood of it, expressed as a product, and let ϕ' be the corresponding modification of type $r + 1$ shrinking S^{r+1}. Note that if S^{n-r-1} is expressed as the union of two cells E_1^{n-r-1} and E_2^{n-r-1}, the former containing p, then B' can be expressed as the union of the neighborhood $E^{r+1} \times E_1^{n-r-1}$ of the cell $E^{r+1} \times \{p\}$ in $E^{r+1} \times S^{n-r-1}$ and a neighborhood of the cell E_0^{r+1} in $M - B$. The latter neighborhood is itself an n-cell E_0^n.

Now consider the construction of the combined trace of the sequence of modifications ϕ followed by ϕ'. It will be remembered that the trace of ϕ is obtained from $(M - B) \times I$ by adding an $(n + 1)$-cell E, with suitable identification of boundary points. The intersection of E with M is B, while its intersection with M' is the set $E^{r+1} \times S^{n-r-1}$ added to $M - B$ in the construction of ϕ. On the other hand, the trace of ϕ' is obtained from $(M' - B') \times I$ by adding an $(n + 1)$-cell E' whose intersection with M' is B'. Thus the intersection of E and E' is the intersection of B' with the set $E^{r+1} \times S^{n-r-1}$. This intersection can be written as $E^{r+1} \times E_1^{n-r-1}$ (cf. the remark made earlier on the construction of B'). The point is that this is an n-cell. Thus the $(n + 1)$-cells E and E' intersect in an n-cell that lies on the boundary of each. It follows that $E \cup E'$ is an $(n + 1)$-cell.

Some more pieces are to be added now to $E \cup E'$, so that the result is still an $(n + 1)$-cell. First, add the set $E_0^n \times I$ in $(M - B) \times I$, where E_0^n is the neighborhood of E_0^{r+1} in $M - B$, as introduced earlier.

This set meets $E \cup E'$ in the union of two sets, namely, $E_0^n \times \{1\}$ (which is its intersection with E') and a set of the form $S^r \times E_1^{n-r-1} \times I$ (which is its intersection with E; this arises since the intersection of E_0^n with B is of the form $S^r \times E_1^{n-r-1}$). It is then easy to see that the union of this set $E_0^n \times I$ with $E \cup E'$ is still an $(n + 1)$-cell.

As an indication of the method of proof of this last statement, note that whenever E^{n+1} is an $(n + 1)$-cell with boundary sphere S^n, then for any n-manifold K with boundary $K \subset S^n$, the union of E^{n+1} and $K \times I$, the points of K being identified with those of $K \times \{0\}$, is an $(n + 1)$-cell. This is almost trivial. The present situation is a little more difficult: there is a subset L of the boundary of K such that S^n contains a set $L \times I$ with $L \times \{0\}$ identified with L, $L \times I$ being otherwise disjoint from K, and this set $L \times I$ in S^n is identified with $L \times I$ in $K \times I$. And it has to be shown that the resulting union is still an $(n + 1)$-cell.

Now it will be noticed that the points added to $E \cup E'$ as just described are all in the trace of ϕ. It will also be noticed that there is a sort of symmetry between ϕ' and ϕ in reverse. That is, ϕ and ϕ' shrink S^{r+1} and S^{n-r-1} (in M'), respectively, the latter being spheres with just one point in common. The further points to be added to $E \cup E'$ are in the trace of ϕ', and bear the same relation to ϕ' in reverse as the points $E_0^n \times I$ already added bear to ϕ. That is, a set of the form $E_1^n \times I$ is to be added, where E_1^n is a neighborhood in M' of E_2^{n-r-1} (the second half of S^{n-r-1}) and boundary points are identified so that the resulting union is still an $(n + 1)$-cell.

Let E'' denote the $(n+1)$-cell obtained by adding sets as just described to $E \cup E'$. Consider now what is left of the traces of ϕ and ϕ' when E'' is removed. What is left of the trace of ϕ is $(M - B) \times I$ with the set $E_0^n \times I$ removed. This is a set of the form $N \times I$, where $N = M - B - E_0^n$. Note that $N \times \{1\}$ is then the complement in M' of neighborhoods of S^{n-r-1} and S^{r+1}. Similarly, the complement of E'' in the trace of ϕ' is a product $N' \times I$. However, since this has the same intersection with M' as $N \times I$, it follows that $N = N'$. Thus the complement of E'' in the combined trace of ϕ and ϕ' is of the form $N \times I$. Here, if I is expressed as the union of the two half intervals I_1 and I_2, $N \times I_1$ will be in the trace of ϕ and $N \times I_2$ in that of ϕ'. In addition, the product structure of $N \times I$ induces a product structure on part of the boundary of E'', while the remainder of this boundary consists of n-cells, namely, $B \cup E_0^n$ in the initial M and a similar set in the result of ϕ followed by ϕ' (which is also M). This product structure can then be extended to an expression of E'' as a product of I and an n-cell, and when this is put together with $N \times I$ it will turn out that the trace of the sequence ϕ followed by ϕ' is expressed as a product $M \times I$. In particular this means that the result of ϕ followed by ϕ' is the identity transformation of M.

The two modifications ϕ and ϕ', related as described in the foregoing, will be called complementary. Thus the result obtained can be summed up by saying that the composition of a pair of complementary modifications is the identity transformation on M and the combined trace is $M \times I$.

8-3. A THEOREM ON 3-MANIFOLDS

The result of the last section has an interesting consequence concerning the structure of an oriented three-dimensional manifold.

In the first place, note that in general, if ϕ is a spherical modification of type r on M satisfying the condition of the last section, then ϕ transforms M into M' and the complementary modification ϕ' transforms M' back into M. But this means that ϕ' in reverse transforms M into M', and ϕ' in reverse is of type $n - r - 2$. That is, if M is transformed into M' by a modification of type r satisfying the condition of the last section, then M can also be transformed into M' by a modification of type $n - r - 2$.

Now let M be a compact orientable three-dimensional differentiable manifold. There is a theorem that says that such a manifold is always the boundary of an orientable 4-manifold (cf. [4]). A 4-cell can be removed from this 4-manifold, leaving a manifold whose boundary is the disjoint union of M and a 3-sphere. Hence (Theorem 6-3), M can be obtained from a 3-sphere by a finite sequence of modifications, and those of types 0 and 2 will be of the orientable kind. Now a modification of type 0 of the orientable kind certainly satisfies the condition of the last section that the sphere S^0 shrunk in such a modification is the boundary of a cell E^1 that closes up to form a directly embedded 1-sphere. Hence, by the remark made earlier, each modification of type 0 on the way from S^3 to M can be replaced by a modification of type 1. The modifications of type 2 are of type 0 in reverse, and so they can also be replaced by modifications of type 1 (here the reverse of a modification of type 1 is also of type 1). Hence, the given M is obtained from the 3-sphere by a finite sequence of modifications, all of type 1.

Stating explicitly what a modification of type 1 actually does leads to the following theorem.

Theorem 8-1. *An orientable compact three-dimensional manifold can be obtained from a 3-sphere by cutting out a finite number of disjoint solid tori (sets of the form $S^1 \times E^2$) and filling the holes again with solid tori, with some suitable identification of boundaries.*

The point here is that the boundary of each hole is of the form $S^1 \times S^1$, and there are in fact infinitely many ways of identifying this with the boundary of a solid torus, and so there are infinitely many different constructions of this kind. It is, however, not easy to decide when two of these apparently different constructions give the same result. In the first place, to do this we would need a solution of the Poincaré problem, which asks whether an orientable 3-manifold in which every 1-sphere can be shrunk to a point is in fact a 3-sphere.

Bibliography

REFERENCES CITED IN THE TEXT

[1] D. W. Blackett, *Elementary topology,* Academic Press, New York (1967).

[2] N. Bourbaki, *Topologie générale,* Hermann, Paris.

[3] R. Courant, *Differential and integral calculus,* Wiley (Interscience), New York (1937).

[4] F. Hirzebruch, *Neue topologische Methoden in der algebraischen Geometrie* (Ergeb. Math. **9**, N. S.).

[5] E. L. Ince, *Ordinary differential equations,* Dover, New York (1953).

[6] J. Munkres, *Elementary differential topology* (Ann. of Math. Stud. **54**) (1963).

[7] S. Smale, *Generalized Poincaré conjecture in higher dimensions,* Bull. Amer. Math. Soc. **66** (1960), 373–375.

[8] A. H. Wallace, *Modifications and cobounding manifolds II,* J. Math. Mech. **10** (1961), 773–809.

[9] A. H. Wallace, *Introduction to algebraic topology,* Macmillan (Pergamon), New York (1957).

GUIDE TO FURTHER READING

The following references are given as a guide to the student who may wish to enter more deeply into the field of differential topology.

The most fundamental concept of the subject is the notion of differentiable structure on a manifold. In the case of a space like the n-sphere we have, so to speak, ready-made local coordinate systems that are fitted together by differentiable transition functions.

But could coordinate systems be defined on the n-sphere in a second way, so that the resulting differentiable manifold would not be diffeomorphic to the first? And is it possible that a space could be topologically a manifold (that is, be covered by neighborhoods that are cells), but not be capable of being made into a differentiable manifold? It has recently been shown that the answers to both of these questions are affirmative. For an introduction to the study of such questions see

J. Munkres, *Elementary differential topology* (Ann. of Math. Study **54**, 1963).

Chapters 7 and 8 indicate that the study of critical points and of spherical modifications should lead to information on the structure of manifolds. Some knowledge of algebraic topology is needed for such a study; given this prerequisite, an introduction to questions of structure of manifolds will be found in

S. Smale, *Generalized Poincaré conjecture in dimensions > 4,* Ann. of Math. **74** (1961), 391–406.

A. H. Wallace, *Modifications and cobounding manifolds,* Pt. I, Canad. J. Math. **12** (1960), 503–528; Pt. II, J. Math. Mech. **10** (1961), 773–809; Pt. III, J. Math. Mech. **11** (1962), 979–990; Pt. IV, J. Math. Mech. **12** (1963), 445–484.

A deep study of classification problems, using similar techniques, will be found in

C. T. C. Wall, *Classifications problems in differentiable topology,* Topology **2** (1963), 253–281, **3** (1965), 291–304, **5** (1966), 73–94.

Finally, the notion of critical points has important applications in differential geometry. For an introduction to this topic see

J. Milnor, *Morse theory,* (Ann. of Math. Study **51,** 1963).

H. Seifert and W. Threlfall, *Variationsrechnung im Grossen,* Chelsea, Bronx, New York (1934).

Index